Methyl Chloroform and Trichloroethylene in the Environment

Authors:

Domingo M. Aviado
University of Pennsylvania School of Medicine
Philadelphia

Samir Zakhari
University of Pennsylvania School of Medicine
Philadelphia

Joseph A. Simaan
American University of Beirut
Beirut, Lebanon

Andrew G. Ulsamer
Bureau of Biomedical Science
United States Consumer Product Safety Commission
Washington, D.C.

Editor

Leon Golberg
Editor-in-Chief
Solvents in the Environment Series
Chemical Industry Institute of Toxicology
Research Triangle Park, North Carolina

Published by

CRC PRESS, Inc.
18901 Cranwood Parkway · Cleveland, Ohio 44128

Library of Congress Cataloging in Publication Data

Main entry under title:

Methyl chloroform and trichloroethylene in the environment.

At head of title: CRC.
Includes bibliographies and indexes.
1. Trichloroethane – Toxicology. 2. Trichloroethylene – Toxicology. I. Aviado, Domingo M. II. Chemical Rubber Company, Cleveland. III. Title: CRC methyl chloroform and trichloroethylene in the environment.
RA1242.T66M47 615.9'51'5 76-16138
ISBN 0-87819-098-8

International Standard Book Number 0-87819-098-8

Library of Congress Card Number 76-16138
Printed in the United States

Direct all inquiries and correspondence to
CRC PRESS, INC., 18901 Cranwood Parkway, Cleveland, Ohio 44128

THE AUTHORS

Domingo M. Aviado is the Professor of Pharmacology and Toxicology at the University of Pennsylvania School of Medicine at Philadelphia. He obtained his M.D. degree from the same University in 1948 and immediately joined the staff of the Department of Pharmacology.

Samir Zakhari is a Postdoctoral Fellow in Pharmacology at the University of Pennsylvania. He obtained his Ph.D. from the Slovak Academy of Science and is on leave of absence from the Pharmacology Unit, National Research Center of Cairo.

Joseph A. Simaan obtained his M.D. degree from the American University of Beirut and is presently an Associate Professor of Pharmacology in the same institution. From September 1974 to August 1975, he was a Visiting Associate Professor of Pharmacology at the University of Pennsylvania.

Andrew G. Ulsamer obtained his Ph.D. degree from Albany Medical College. He is presently the Supervisory Research Chemist, Bureau of Biomedical Science of the United States Consumer Product Safety Commission, Washington, D.C.

This monograph is dedicated to Harold C. Hodge, Ph.D.

FOREWORD

To introduce to the reader a series of volumes on the theme of "Solvents in the Environment" is both a pleasure and a privilege. The approach to the material presented in this series is an innovative one, marking a distinct departure from previous reviews of the toxicology of solvents. In the present and subsequent volumes, critical coverage of the literature will be combined with the fruits of original and highly relevant research carried out by Dr. D. M. Aviado and his colleagues. The total effort constitutes an authoritative up-to-date appraisal of the knowledge needed to understand the inhalational toxicity and other biological effects of an important and ubiquitous class of chemical agents.

The sense of privilege is heightened by the dedication of the first volume in this series to one of the greatest living pioneers in the fields of Toxicology and Pharmacology, Professor Harold C. Hodge. Having long been familiar with the work of this world-famous scientist, it was a pleasant surprise to meet him in person one day in 1962, when a quiet, unassuming man dropped by unannounced, in our offices at the British Industrial Biological Research Association in London's Regent Street. At that time he was characteristically in the forefront of the battle over fluoridation of water supplies and his presentation at the British Nutrition Society was a masterpiece of toxicological work and thought. At BIBRA's first Annual Scientific Meeting he was the principal speaker and his address (*Food Cosmet. Toxicol.,* 1, 25, 1963) still reflects current research needs in the toxicology of food additives. Subsequently we at BIBRA had occasion to review and condense a large volume of Hodge's unpublished safety data on phosphates, in our series entitled, "Summaries of Toxicological Data" (*Food Cosmet Toxicol.,* 2, 147, 1964). The links thus established have been maintained over the past 14 years through Professor Hodge's membership of the Honorary Advisory Board of *Food and Cosmetics Toxicology*.

It is appropriate to recognize Dr. Hodge's contributions towards the objectives represented by the present series of monographs on "Solvents in the Environment." He has been able to demonstrate in an effective manner, over a wide range of chemical agents, and in the course of activities spanning many years, the strategic deployment of pharmacologic and toxicologic approaches to the assessment of hazard and the evaluation of safety. The results were always weighed with care and objectivity. Scientific expertise, and not emotion, was the basis of the judgments arrived at. If the new generation of toxicologists and regulatory officials can follow in these footsteps, our society will be the beneficiary of his labors.

Any work on the toxicology of solvents must fall short of perfection to the extent that it does not cover what at first sight might appear to be technological and industrial aspects such as specifications, characteristic impurities and their levels, and the contributions made by them to the overall toxicity of the particular product of commerce. The nature and levels of associated compounds, such as stabilizers, may also influence the biological properties of a solvent. Unfortunately, major gaps exist in our knowledge of these areas of toxicology. This fact, more than any other, sets a limit to the size of each monograph and to the encyclopedic breadth of coverage that might theoretically be possible. Readily available sources and reference manuals provide the technical data to supplement the information in these monographs. The object here is to distil and condense, making available to a wide range of potential readers the biological data that they will find most useful. The editor and the authors are confident that the objective has been fulfilled.

Leon Golberg
Research Triangle Park, North Carolina

ACKNOWLEDGMENTS

We would like to thank Alfredo Bianchi and Charles L. Weissman for their assistance in the preparation of Chapters 3 and 8. The original investigation described in Chapters 3, 4, 8, and 9 has been funded at least in part with federal funds from the United States Consumer Product Safety Commission under contract number CPSC-C-75-078.

The content of this monograph does not necessarily reflect the views of the Commission, nor does mention of trade names, commercial products, or organizations imply endorsement by the Commission.

TABLE OF CONTENTS

Part I
Methyl Chloroform

Chapter 1

INTRODUCTION TO METHYL CHLOROFORM

In recent years, aerosol products have received a considerable amount of attention. The increase in commercial production of aerosol units has been phenomenal during the past 15 years because of their convenience to the user. The consumption of the aerosol products would have continued to increase were it not for the questions raised as to the possible toxicity of the aerosol products both directly to the consumer or indirectly to the entire population. The direct effects of aerosols on humans are the immediate concern of this review. Indirect effects such as the possible effects of fluorocarbon propellants on the stratosphere, resulting in an increase in incidence of skin cancer, are not within the scope of this monograph.

One of the authors (D. M. A.), in 1970, initiated an investigation of aerosols as they relate to the treatment of bronchial asthma in general and to the use of sympathomimetic bronchodilators in particular.[1] At that time, the aerosols were widely used to dispense bronchodilator drugs and it became immediately apparent that there was no published information on the toxicity of such aerosols.[2,3] There were scattered reports of deaths among asthmatics attributed to aerosol products containing bronchodilators and the literature has been reviewed elsewhere.[4] Although the fluorocarbon propellants used in aerosols were originally thought to be inert, there is now definite evidence that they are toxic, especially at higher concentrations, and that the most widely used propellants are the most hazardous.[5,6] After the effects of 15 propellants were investigated in four animal species, a system of classification was introduced, grouping the high-pressure and low-pressure propellants according to high, intermediate, and low levels of toxicity.[7-18]

The classification of the aerosol propellants on the basis of acute inhalational experiments was the initial step in the understanding of toxicity of aerosol products. Currently the authors and their collaborators are examining the other constituents of aerosol products used as drugs (such as bronchodilators and medicinal vapors), as cosmetics, and as household products. The plan is to publish the results of the investigation as they relate to the body of available information on a particular aerosol product, a specific propellant or a selected solvent. This initial report is on methyl chloroform, which is used as a commercial solvent in industry and is contained in consumer products.

Chapter 2

REVIEW OF THE LITERATURE ON METHYL CHLOROFORM

Although methyl chloroform was synthesized as early as 1887, its use in aerosol consumer products did not start until about 1957. As soon as it was produced on a large scale, a commercial firm[19] introduced it as a general solvent with the justifiable claim that methyl chloroform was less toxic than chloroform and carbon tetrachloride. The literature on the toxicity of methyl chloroform has been reviewed by Stewart,[20-23] Blankenship,[24] and Chenoweth and Hake.[25] There is limited information relating to the effect of the solvent when inhaled as an aerosol. However, it is possible to interpolate to some extent data from the literature on the potential hazards of exposure to methyl chloroform in aerosols. In this monograph, it is intended to examine the publications on human and animal exposures to methyl chloroform. The experiments that are described in Chapters 3 and 4 are based on the toxicity reported in the literature reviewed in this section.

A. Chemical and Physical Properties

Synonyms: methyl chloroform; alpha-trichloroethane.

Chemical name: 1,1,1-Trichloroethane (CH_3CCl_3); C = 18.00%, H = 2.27%, Cl = 79.72%.

Trade names: Aerothene TT; Axothene No. 3; Barcothene Nu; Blakeothane; Blakesolv 421; CF2 Film Clean; Chloromane; Chlorothene Nu; Chlorothene VG; Chlorothene; Dowclene WR; Dyno-Sol; ECCO 1550; Ethyl 111 Trichloroethane (MPG); Insolv Nu; Insolv VG; Kold Phil; Kwik-Solv; Lectrosolv 170; Methyl Chloroform Tech; Nacon 425; One, One, One; Penolene 643; Perm-Ethane DG (Permathane); Saf-T-Chlor; Solvent M-50; Solvent 111; Sumco 33; Tri-Ethane; Tri-Ethane Type 314; Tri-Ethane Type 315; Tri-Ethane Type 324; Tri-Ethane Type 339; Triple One; V-301; Vatron 111.

Molecular weight: 133.42.

Boiling point: 74.1°C; solidifies at −32.5°C; refractive index 1.43765 (21°C).

Specific gravity: 1.336 at 25°C; viscosity (25°C); 0.79 centipoise.

Vapor pressure: 127 mm Hg at 25°C; 103 mm Hg at 20°C.

Vapor density: 4.62 (air = 1); at room temperature air is saturated at a concentration of 160,000 ppm (0.9 g/l).

Soluble: in acetone, benzene, carbon tetrachloride, methanol, ether, carbon disulfide; in water 260 ppm at 25°C.

Colorless liquid possessing a distinctive chloroform-like odor, but is not particularly disagreeable or irritating in 1 to 2% concentration.

Thermal Stability: Considerable amounts of hydrogen chloride are obtained from methyl chloroform passed over iron, steel, or copper at 282°C. However, only trace quantities of phosgene ($COCl_2$) are formed above 315°C, e.g., 1 g methyl chloroform produces 0.74 mg of phosgene when passed over iron at 335°C.[26] Hydrogen chloride is produced more readily than phosgene owing to the presence of hydrogen atoms in the methyl chloroform molecule.[27] In order to promote its stability, the agent is usually stored in dark brown bottles and inhibited against oxidation and light degradation. Like many chlorinated hydrocarbons, methyl chloroform reacts with aluminum alloys and must be inhibited if corrosion is to be prevented. It is noncorrosive when used in anhydrous aerosol formulation.[28]

Flammability: nonflammable; does not support combustion in air at standard temperature and pressure; limits of flammability of the vapors of inhibited compounds are 10 to 15.5% in air with hot wire ignition only; no flash point or fire point using standard ASTM procedures.

Preparation of methyl chloroform: by the substitution of chlorine into 1,1-dichloroethane; or by the catalytic addition of hydrogen chloride to 1,1-dichloroethylene.

Determination of methyl chloroform in biological fluids: The modified method of Parker et al.[29] in screening for the presence of volatile organic materials is routinely used. This procedure depends on the gas chromatographic determination and detects numerous volatile substances when present in concentrations of over 20 mg%. The details of the detection of methyl chloroform in the blood are described by Hall and Hine.[30]

B. Industrial Uses

Methyl chloroform has been used as a solvent primarily for vapor degreasing, cold-cleaning, dip-cleaning, and bucket-cleaning of metal for the removal of greases, oils, and waxes. Historically, it has been a common substitute for carbon tetrachloride. Among other industrial uses are cold-cleaning and vapor degreasing in the aircraft, automotive, electronic, and missile industries; maintenance and cleaning of motors, appliances, and equipment; and on-the-site cleaning of printing presses, food packaging machinery, and molds.

C. Aerosol Products

Methyl chloroform is used in aerosol formulations, both as a solvent and as a low-pressure propellant. When it is mixed with a high-pressure propellant, there is a reduction of pressure in the aerosol container and thereby an increase in the safety of the product.[31] The aerosol products containing methyl chloroform are as follows:

1. Hair Sprays and Cosmetics

Schober[19] reviewed the use of methyl chloroform in aerosol products. He pointed out the advantages as "low cost, low toxicity, low fire hazard and high quality in the final product." It is further claimed that "extensive toxicological studies were performed on the solvent and on hair sprays containing methyl chloroform." This article appeared in 1958 and nothing more on toxicity of the aerosol product has been reported since that time.

In 1967, a propellant system containing chlorinated solvents and a soluble compressed gas was introduced for use in aerosol products.[24,28] The combination of methyl chloroform and methylene chloride (trade name Aerothene) was used in combination with nitrous oxide or carbon dioxide for generating pressure. It is claimed that their excellent solvency, rapid dry rate, low toxicity, and optimum protection against can corrosion are among their desirable features.[24] There has been no report of toxicologic investigation of the aerosol generated by this system.

2. Medicated Vapors

In 1973, 21 deaths resulted from the abuse of decongestant aerosol sprays containing methyl chloroform[32,33] and fluorocarbon propellants. The Food and Drug Administration has issued an order requiring manufacturers to register drugs for human use containing methyl chloroform.

3. Consumer Aerosol Products

It has been difficult for the author to obtain the names of aerosol products containing methyl chloroform directly from the manufacturers. Gleason et al.[34] have listed 14 products which are used as follows:

cleaners:
- oven cleaner – one product (65% methyl chloroform)
- spot removers – five products (concentrations of methyl chloroform known in three products: 25%, 60% and 70%)

waxes and polishes:
- furniture polish – two products

automotive products:
- lubricant – one product (70%methyl chloroform)
- choke cleaner – one product
- degreaser – one product

specialty products:
- glass water repellent – one product (88% methyl chloroform)
- suede water proofing – one product (42% methyl chloroform)
- adhesive – one product

The above list is based on information accumulated by Gleason et al.[34] According to the same source, there are 45 nonaerosol products containing methyl chloroform in concentrations ranging from 10 to 100%.

D. Human Investigation

There is no investigation directly relating to the health effects of normal usage of aerosols containing methyl chloroform. There is information on lethal exposure as well as on experimental inhalation of low concentrations of methyl chloroform.

1. Fatalities in Confined Space

There have been 11 reported fatalities occurring in solvent tanks. The first death reported by Kleinfeld and Feiner[35] was that of a public utility worker who entered an underground vault (14 X 7 X 7 feet) which served as a distribution point for a substation, to remove accumulated grease from conduits. Rags dipped in methyl chloroform were used for the purpose. The worker experienced symptoms of giddiness and light-headedness and then collapsed after 10 minutes' exposure in the vault. It was estimated that the concentration within the vault was 27.5 mg/l. The authors reported that "Toxicological analysis showed excessive concentrations of methyl chloroform in the blood, kidney, brain, and liver," but did not give the amounts. There were three other deaths resulting from overexposure in confined spaces, but the details were not reported.[22,36]

The next six fatal cases of methyl chloroform poisoning were obtained from files at the Armed Forces Institute of Pathology by Stahl et al.[37] The reported deaths were from exposure to methyl chloroform in enclosed spaces. Analysis of tissues showed varying levels of methyl chloroform. In the brain, the concentration in mg/100 g tissue for the six cases were as follows: 0.32, 2.7, 9.3, 50.0, 56.0, and 59.0.

The eleventh fatality from methyl chloroform exposure was reported by Hatfield and Maykoski.[38] A radiator and metal tank repairman was found by his fellow workers in the tail portion of a 450-gallon aircraft tip tank with only his legs protruding from the upper end of the tank. One hour and 20 minutes after the discovery of the body, a concentration of methyl chloroform of 2.75 mg/l was detected in the tank. A reconstruction of the incident indicated that saturation of cleansing pads may have caused the concentration to reach 341 mg/l and that one hour later it may have been 242 mg/l. The lethal concentration was estimated to be between 341 and 2.75 mg/l.

2. Nonindustrial Fatalities

In addition to the 21 deaths from abuse of decongestant aerosols,[32,33] there were two cases of abuse of methyl chloroform reported by Hall and Hine.[30] Both involved inhalation of cleansing solvents; the first was a case of chemical pneumonia resulting from aspiration and the other involved death from respiratory arrest. The respective blood levels of methyl chloroform were 13.0 and 72 mg/100 ml. The third case of nonindustrial poisoning was described by Stewart and Andrews[39] and consisted of accidental ingestion of 0.6 g/kg of methyl chloroform. There were signs of depression of the central nervous system. The patient recovered after supportive treatment.

The fourth and most recent fatality from methyl chloroform was reported in 1974 by Travers.[40] An 18-year-old apprentice seaman collapsed on the deck of his ship. A bottle of methyl chloroform was found in his bunk along with a rag soaked with the solvent. After different supportive measures were taken, progressive hypotension and bradycardia, unresponsive to isoproterenol and norepinephrine infusions, and several episodes of cardiac arrest eventuated in his death 24 hr after collapse. Autopsy showed right atrial and ventricular dilation and circumferential left ventricular subendocardial hemorrhage. Microscopically, widespread recent infarction was observed. Mild congestion of viscera, cerebral edema, and Purkinje cell chromatolysis were also noted. Postmortem analysis failed to detect methyl chloroform in the liver, kidney, blood, or brain.

3. Experimental Inhalation

The experimental exposure of human subjects to methyl chloroform in a chamber was first reported by Torkelson et al. in 1958.[36] Single exposures caused the following:

a. 2.48 to 3.9 mg/l for 90 min: No effect other than the smell of the chemical.

b. 2.28 to 3.25 mg/l for 450 min: Moderate smell, which tended to disappear. Examination made before and after exposure indicated no significant changes in pulse, respiration, blood pressure, reflexes, or equilibrium. Tests for urinary urobilinogen, thymol turbidity, and excretion of bromsulphalein did not indicate any significant changes in liver function.

c. 4.9 to 6.55 mg/l for 30 min: Unpleasant odor; equilibrium apparently not disturbed.

d. 4.95 to 5.5 mg/l for 75 min: Two of the four individuals exposed reported a strong odor and one reported a slight eye irritation. Three of the four reported a feeling of light-headedness. The Flanagan tests given during exposure and the Romberg tests given immediately after exposure revealed a slight but definite loss of coordination and equilibrium. Recovery occurred quite rapidly within 5 to 10 min. All electrocardiograms were normal throughout the exposure. No significant changes were found in the following tests and determinations: Cephalin flocculation, thymol turbidity, urinary urobilinogen, complete urinalysis, serum iron, serum glutamin-pyruvic transaminase, hemoglobin, sedimentation rate, and red, white, and differential blood counts.

e. 9.57 to 11.99 mg/l for 5 min: Very noticeable odor. Obvious disturbances of equilibrium. Romberg test positive.

A series of six controlled human exposures to methyl chloroform was conducted by Stewart et al. in 1961.[41] The subjective and physiological responses to a constantly increasing vapor concentration over a 15-min period were as follows:

a. 0 to 5.5 mg/l for 15 min: Increasing awareness of a slightly sweet, not unpleasant odor.

b. 5.5 to 6.05 mg/l for 15 min: Mild eye irritation noted in six of seven subjects.

c. 10.45 to 11.0 mg/l for 15 min: Throat irritation noted in six of seven subjects.

d. 14.3 mg/l for 15 min: One subject very light-headed.

e. 14.58 mg/l for 15 min: Two subjects unable to stand; three subjects very light-headed but able to stand; two subjects were not light-headed and one of them was able to demonstrate a normal Romberg test.

The presence of methyl chloroform in blood during the exposures and the concentration of the compound in the expired air after the exposures were determined by infrared spectrometric methods. Methyl chloroform was observed to give a prolonged exponential decay curve when its concentration in the postexposure expired air was plotted versus hours of postexposure. This decay curve was predictable enough to permit a reasonable estimation of the magnitude of exposure, 0.5 to 3 hr after the exposure had occurred. Therefore, the analysis of postexposure expired air allowed a positive identification of the compound to be made and provided data with which the magnitude of the exposure might be estimated.

Clinical laboratory studies indicated that the determination of the concentration of urinary urobilinogen proved to be the most sensitive test of liver stress produced by markedly excessive exposures to methyl chloroform. The concentration of 2.75 mg/l was found to be a satisfactory threshold limit for safety of methyl chloroform.[41]

Eleven human subjects were experimentally exposed to methyl chloroform vapor (2.75 mg/l) for periods of 6.5 to 7 hr per day for 5 days. No clinical laboratory test performed during or following the vapor exposure revealed any abnormality of organ function. The only adverse objective response was an abnormal modified Romberg's test observed in two of the subjects during exposure.[42]

In the inhalational exposure described above, performed by Stewart et al.,[41] the lowest concentration was 2.75 mg/l for 78 minutes. A more extended period was used by Salvini et al.[43] They exposed six students to an average vapor concentration of 2.48 mg/l for two periods of 4 hr, separated by a 1 1/2 hr interval. The psychophysiological functions studied were not decreased. No disturbance in motor function, coordination, equilibrium, or behavior patterns was observed. The low percentage reduction in performance which was observed is not statistically significant. In the perception test an interaction between exposure to methyl chloroform and mental fatigue was observed.

A fifth investigation was conducted by Gazzaniga et al.[44] who exposed six volunteers for 3 hr in a chamber containing 3.26 mg/l of methyl chloroform. All the subjects tolerated the concentration without manifesting either subjective effects or alteration in respiration and pulse rate.

The sixth group of observations actually predated those reported above. In 1960, Dornette and Jones[45] attempted to induce general anesthesia in 50 patients by administration of methyl chloroform in the following concentrations: 1.0 to 2.6% (55.5 to 144.3 mg/l) for induction; 0.6 to 2.25% (33.3 to 124.9 mg/l) for maintenance of light anesthesia for up to 2 hr. In all instances, a combination of 80% nitrous oxide and 20% oxygen was administered. Respiration seemed neither depressed nor stimulated by the inhalation of methyl chloroform. Monitoring of the electrocardiogram during 32 administrations revealed the following arrhythmias:

a. nodal rhythm: six patients.

b. occasional premature ventricular contractions: three patients. It was noted that the development of premature ventricular contractions was always associated with respiratory obstruction. Reversal to normal rhythm could be accomplished by relieving the obstruction and adequately ventilating the patient's respiratory tract.

c. frequent premature ventricular contractions: two patients.

d. depressed S-T segment: two patients.

There was a fall in blood pressure of 5 to 10 mm Hg systolic in about half of the patients; in three patients the systolic pressure fell to a greater degree and was 60 mm in one patient. In all three the blood pressure returned to the preoperative range when the concentration of the agent was decreased and the level of anesthesia lightened.

4. Inspired and Expired Concentrations

The techniques for measurement of inspired concentrations of methyl chloroform were reviewed by Simmons and Moss.[46] The measurement depends on trapping methyl chloroform vapors on silica gel contained in a robust sampling tube made from a galvanized iron pipe connector and Simplifix couplings. Absorbed solvent is then eluted with methanol and estimated by gas chromatography, Chromosorb 101 being used as a stationary phase. The analysis of expired air is simple, rapid, and sensitive, and is practical for clinical use. The amount of solvent in the breath is dependent upon the total amount absorbed by the body, the length of time over which the solvent was absorbed, and the time elapsed since the exposure. Hence, under standardized conditions each solvent possesses a characteristic excretion or decay curve in breath. This has been applied to the following problems:

a. diagnosis of poisoning;[39]

b. estimation of magnitude of exposure by inhalation;[41]

c. estimation of rate of excretion;[47,48]

For the last-mentioned application, Morgan et al.[49,50] used radioactive chlorine-38 labeled halogenated hydrocarbons and the excretion in breath was followed for 1 hr. Methyl chloroform was absorbed more rapidly and excreted more rapidly than fluorocarbons. It was noted that the partition coefficient between blood and air correlated with the retention pattern. The significance of the partition coefficient in relation to the anesthetic action and to the relative toxicity of chlorinated hydrocarbon was reviewed by Morgan et al.[49]

5. Urinary Excretion

The determination of urinary urobilinogen proved to be the most sensitive test of liver stress produced by exposure to methyl chloroform. Stewart et al.[41] reported that urine collected after 15 min exposure to concentrations of up to 14.58 mg/l of methyl chloroform did not contain measurable amounts of the solvent. However, it caused a decrease in urobilinogen levels; the pre-exposure level was 1:10 and the 6-hr post-exposure level was 1:640.

All other investigations of urinary excretion in man have dealt with the identification of the metabolites of methyl chloroform. Tada et al.[51] exposed two male volunteers to methyl chloroform and reported an increase in urinary excretion of trichloroacetic acid, as determined by the alkaline pyridine method. However, the increase was not in proportion to the concentration of the vapor and the duration of exposure, as far as the exposures to 1.16 and 2.26 mg/l were concerned. Monzani et al.[52] reported that among 18 workers exposed to 1.0 mg/l, only one showed a detectable amount of trichloroacetic acid (9.72 mg/l urine). Seki et al.[53] examined the urine samples collected from seven workers and on the basis of excretion of trichloromethanol estimated the biological half life to be 8.7 hr for methyl chloroform.

E. Absorption, Metabolism, and Disposition in Animals

Methyl chloroform is rapidly absorbed through the lungs and the gastrointestinal tract. It may be absorbed in toxic quantities through the intact skin if trapped against the skin beneath an impermeable barrier. Following absorption, most of the compound is eliminated unchanged via the lungs. Hake et al.[54] estimated that nearly 98% was eliminated unchanged in the expired air of the rat, following an intraperitoneal injection of carbon-14 labeled compound. One-half percent of the dose was metabolized to carbon dioxide, while the remainder appeared in the urine as the glucuronide. The animal breath excretion data have been used by Boettner and Muranko[55] for estimation of total amount of exposure of humans to chlorinated hydrocarbons by postexposural analysis of the breath.

Subsequent work has identified another metabolite in the urine. Ikeda and Ohtsuji[56] exposed a group of rats to 1.1 mg/l of methyl chloroform for 8 hr. The urine collected for 48 hr from the beginning of exposure was analyzed and the results were as follows:

Total trichlorocompounds	3.6 ± 1.0 mg/kg
Trichloroacetic acid	0.5 ± 0.2 mg/kg
Trichloroethanol	3.1 ± 1.0 mg/kg

After intraperitoneal injection of 0.37 g/kg the results were as follows:

	First 48-hr period	Second 48-hr period
Total trichlorocompounds	4.0 ± 1.5 mg/kg	0.3 ± 0.1 mg/kg
Trichloroacetic acid	0.5 ± 0.2 mg/kg	0.3 ± 0.1 mg/kg
Trichloroethanol	3.5 ± 1.4 mg/kg	0 mg/kg

It appears from the above that urinary excretion can be used to estimate total uptake of methyl chloroform, although the amount is small in proportion to the amount exhaled. Tada et al.[51] found that the amount of trichloroacetic acid was increased by repetition of exposure but not in proportion to the concentration and duration of exposure to methyl chloroform.

Subacute inhalation studies in rats completed by Eben and Kimmerle[57] showed a phenomenon that has not been reported hitherto. A daily 8-hr exposure for 3 months to a concentration of methyl chloroform of about 1.08 mg/l (20 ppm) influenced urinary excretion of trichloroethanol; this excretion increased until the tenth week and then it fell and remained constant at a level somewhat below the maximum concentration. Since there was no accumulation of methyl chloroform in the organs, the authors concluded that methyl chloroform caused an induction of hepatic enzyme which was responsible for the increase in formation of trichloroethanol. The enzyme induction caused by methyl chloroform is discussed further under Section G-4 on hepatotoxicity.

F. Toxicologic Investigation in Animals

The comparisons of lethal dosages for chloroform, carbon tetrachloride, and methyl chloroform established that the last-mentioned compound has the lowest level of toxicity. Of the two isomers of methyl chloroform, 1,1,1-trichloroethane is less toxic than 1,1,2-trichloroethane.[58-60] This review does not cover the latter isomer.

1. Oral Administration

Torkelson et al.[36] determined the lethal dose (LD_{50}) following intragastric administration of methyl chloroform. The results of the uninhibited and inhibited forms of methyl chloroform were as follows:

	Uninhibited	Inhibited
Rats (male)	12.3 g/kg	14.3 g/kg
Rats (female)	10.3 g/kg	11.0 g/kg
Mice (female)	11.24 g/kg	9.7 g/kg
Rabbits (mixed sex)	5.66 g/kg	10.5 g/kg
Guinea pigs (mixed sex)	9.47 g/kg	8.6 g/kg

In four animal species, both forms of methyl chloroform have low levels of oral toxicity and for practical purposes their toxicities are the same.

2. Intraperitoneal Injection

Several investigators have determined the LD_{50} of methyl chloroform as follows:

Mice[60]	2.56 g/kg
Mice (female)[62]	4.7 g/kg
Mice (male)[63]	4.94 g/kg
Rats (male)[64]	5.08 g/kg
Dogs[65]	4.14 g/kg

In the above list, the second and third values for

mice are based on injection of methyl chloroform in corn oil, which retards absorption. If the first value is used, it appears that of the three animal species, mice are more sensitive than rats and dogs to lethal action of methyl chloroform.

3. Inhalation Toxicity: Single Exposure

The first investigation of vapor toxicity of methyl chloroform was reported in 1950 by Adams et al.[60] In rats the following concentrations in single exposures caused 50% of deaths (LC_{50}): 98 mg/l for 3 hr and 76 mg/l for 7 hr. The following exposures were not lethal but caused depression of the central nervous system, ranging from a mild narcotic effect and a decreased tendency of rats to move, to the appearance of stupor and unconsciousness: 98 mg/l for 0.3 hr, 55 mg/l for 3.0 hr, and 44 mg/l for 7.0 hr.

Postmortem examination of rats showed that single exposure at 45 mg/l for 5 hr and at 98 mg/l for 0.3 hr did not produce organic injury. On the other hand, the following single exposures caused pathologic lesions:

a. 98 mg/l for 2 hr produced a slight increase in the weight of the kidneys without any significant pathologic changes.

b. 66 mg/l for 7 hr produced a significant increase in the liver weight with histologic lesions of slight to moderate intensity. Numerous small clear vacuoles were observed microscopically throughout the cytoplasm of the hepatic cells. Considerable congestion and hemorrhagic necrosis occurred in the central area. There was also a slight increase in kidney weight; however, no pathologic changes were observed.

c. 44 mg/l for 7 hr produced fatty changes in the liver of the same type but less severe than those observed in the previous group.

Siegel et al.[66] determined the 4-hr lethal concentration$_{50}$ for methyl chloroform in rats to be 100 mg/l. There are no estimations of LC_{50} for exposure shorter than 4 hr.

4. Inhalation Toxicity: Repeated Exposures

Adams et al.[54] extended their investigation to include subacute toxicity. The results of 7-hr exposures given 5 days a week for 4 to 10 weeks were as follows:

a. 27.3 mg/l in rats caused slight retardation of growth in female but not in male rats during the first 2 weeks; however, the gain in weight for both sexes over the entire experimental period was almost the same as that of the controls. Organ weights, blood urea nitrogen values, and the results of histopathologic examination showed no ill effects.

b. 27.3 mg/l in guinea pigs caused loss of weight during the first 3 weeks and then there was a slow gain, with a final weight increase of 6 and 19% for the females and males, respectively, as compared with 39 and 38% for the controls. All of the exposed guinea pigs showed slight to moderate central fatty degeneration of the liver. In the males varying degrees of testicular degeneration also occurred. The blood urea nitrogen concentration was 25.9 mg/100 ml in exposed guinea pigs, as compared with 28.6 mg/100 ml in the controls.

c. 16.4 mg/l in rats and monkeys caused no adverse effects.

d. 16.4 mg/l in guinea pigs caused a significant depression of growth and definite, although slight, central fatty degeneration of the liver.

e. 8.2 mg/l in guinea pigs caused barely significant retardation of growth but no organic injury.

The weight response of the guinea pigs may represent an unusual species susceptibility, the significance of which is uncertain.

Torkelson et al.[36] examined only the effects of repeated exposures 5 days a week for several months. Their results were as follows:

a. 2.75 mg/l (7 hr per day) for 6 months in rats, guinea pigs, rabbits, and monkeys caused no detectable effect on growth, hematological value, and postmortem examination.

b. 55 mg/l (1 hr or less per day, 5 days per week) for 3 months in rats caused an increase in liver weight. This concentration induced various degrees of anesthesia ranging from ataxia to semiconsciousness.

c. 11 mg/l (1 hr or less per day, 5 days per week) for 3 months in female guinea pigs caused irritation of the lungs and fatty changes in the liver. Guinea pigs exposed for 0.2 of an hour per day were normal except for microscopical evidence of interstitial inflammation of the lungs and a somewhat increased incidence of interstitial nephritis in the kidney.

d. 5.5 mg/l (1 hr or less per day, 5 days per

week) for 3 months in female guinea pigs produced retardation of growth and evidence of inflammation of the lungs, central lobular fatty changes in the liver, and increase in liver weight. Guinea pigs exposed 1.2 hr per day grew normally. Microscopically, however, there was evidence of inflammation or pneumonia in the lungs.

The third inhalational study was performed by Prendergast et al.[67] In a 30-day exposure (8 hr, 5 days a week) at a concentration of 12.1 mg/l, no animals died and no visible signs were noted. Rabbits and dogs showed a body weight loss. Gross and histopathologic examination of brain, heart, lung, liver, spleen, and kidney did not reveal any abnormalities. Serum urea nitrogen determinations in guinea pigs showed no elevation (25 ± 4 mg/100 ml versus 24 ± 5 mg/100 ml for control). Following a 90-day continuous exposure to 2.04 mg/l, there was less body weight gain in the dogs and rabbits than in control animals. One rat had gray nodules on the lower lobe of the left lung and one rabbit had grapelike sacs containing clear fluid on the abdominal wall and adjacent organs. Nonspecific inflammatory changes in the lungs of all species were revealed by microscopy. The results of the 90-day exposure to 0.74 mg/l were difficult to interpret because of the death of some animals. Survivors exhibited varying degrees of lung congestion and pneumonitis. It should be noted that in the higher exposure at 2.04 mg/l no deaths occurred and no pathologic abnormalities were observed.

There are two other publications that should be considered briefly. Tsapko and Rappoport[68] reported effects of acute inhalation in several animal species. Rowe et al.[69] investigated a mixture of 75% methyl chloroform and 25% tetrachloroethylene. They found that animals given 7-hr exposures 5 days a week for 6 months to a vapor concentration of 5.5 mg/l of the mixture suffered changes of a reversible character in the liver and kidney. Human subjects who were exposed to the same mixture, 7 hr a day for up to 5 days to a vapor concentration of 2.75 mg/l, showed no adverse effects.

G. Pharmacodynamic Effects

Like other halogenated hydrocarbons, methyl chloroform influences the functions of the central nervous system, heart, lungs, liver and kidneys. The toxicologic effects of methyl chloroform compared to other related compounds have been reviewed by Llewellyn.[70]

1. General Anesthetic

Von Oettingen reviewed the earlier[71] and subsequent[72,73] literature on the pharmacodynamic effects of methyl chloroform. In the late 1800s this anesthetic was regarded as superior to chloroform because methyl chloroform produces neither excitation nor salivation. Lazarew[74] determined the concentration causing complete narcosis to be 45 mg/l and the minimal fatal concentration 65 mg/l. The ratio between the concentration of the vapor causing the loss of reflexes and that producing death in mice was found to be 20, as compared with 15 for chloroform.

Krantz et al.[75] estimated the concentrations in dogs to be 0.45 g/kg for induction of anesthesia and 0.80 g/kg for causing respiratory failure. The anesthetic index of methyl chloroform was 1.77 for dogs and 2.15 for monkeys, which applications provide greater margins of safety than those of chloroform. The anesthetic use of methyl chloroform in man has been reported by Dornette and Jones[45] and is discussed above. (Page 8.)

The principal toxic action of a single vapor exposure is a functional depression of the central nervous system.[21] Humans exposed to 4.95 to 5.5 mg/l experience transient mild eye irritation and prompt, though minimal, impairment of coordination,[36,41] Above 9.35 mg/l disturbances of equilibrium in humans have been observed.[41] Exposures of this magnitude also may induce headache and lassitude.

2. Cardiotoxicity

The proarrhythmic activity of methyl chloroform has been investigated in the dog. Rennick et al.[76] demonstrated sensitization of the heart to epinephrine-induced arrhythmias after the inhalation of 0.33 to 0.53 g/kg of methyl chloroform in dogs under barbital anesthesia. They also concluded that ventricular arrhythmias could be more regularly induced with methyl chloroform than with chloroform but less than with cyclopropane. Reinhardt et al.[77] found the minimal concentration that causes sensitization in the dog to be 27.8 mg/l. The effective concentration$_{50}$ was 40.7 mg/l in another group of dogs examined by Clark and Tinston.[78]

Somani and Lum[79] and Lucchesi[80] instilled 133.6 mg/kg of methyl chloroform intratracheally and injected epinephrine (10 μg/kg) intravenously. This combination caused ventricular fibrillation except in dogs that were pretreated with beta-adrenergic blocking agents.

In the absence of injection of epinephrine, the administration of methyl chloroform produces alterations in the electrocardiographic pattern as follows:

Humans[45]	nodal rhythm premature ventricular contractions depressed S-T segment
Monkeys[75]	flattened or inverted T wave
Dogs[75]	flattened or inverted T wave
Mice[7,81,82]	2nd degree block and ventricular fibrillation

There is also a depression in contractility of the perfused frog heart,[74] the canine heart-lung preparation,[11] and the primate heart *in situ.*[10] A depression of oxygen consumption also occurs in heart slices obtained from rats anesthetized with methyl chloroform but not in those from unanesthetized rats.[75] As a result of myocardial depression a fall in systemic blood pressure is detected in the dog and the monkey when anesthetized with methyl chloroform.

Herd et al.[83] found that the peripheral vasodilation due to methyl chloroform could be reversed by phenylephrine hydrochloride. Furthermore, injection of calcium ion ameliorated the depression of myocardial contractility and hypotension, induced by methyl chloroform.

3. Pneumotoxicity

Histological examination of the lung of animals exposed to methyl chloroform shows congestion and inflammatory changes (see Section F-4 above). In the monkey, inhalation of 138.8 to 277.5 mg/l of methyl chloroform causes depression of respiratory minute volume accompanied by a decrease in airway resistance and an increase in pulmonary compliance.[14] The pulmonary toxicity of eight chlorinated solvents was studied in rabbits by the physiogram method. The overall toxicity of methyl chloroform was less than the corresponding unsaturated compound trichloroethylene. The presence of a double bond provoked a central depression greater than that observed with the corresponding saturated compounds.[84] No information is available on pneumotoxicity in other animal species.

4. Hepatotoxicity

The effects of methyl chloroform were investigated in several animal species. The mouse has been studied most extensively. The intraperitoneal injection of 5.3 to 16.0 g/kg prolonged the sleeping time induced by pentobarbital, an effect brought about by interference with metabolizing enzymes in the liver; the hepatotoxic effective $dose_{50}$ (ED_{50}) was found to be 11.2 g/kg.[85] However, it is difficult to exclude the possibility that the prolongation of sleeping time is caused by the addition of the hypnotic effects of pentobarbital and methyl chloroform, instead of the latter producing hepatic lesions.

Additional signs of hepatic dysfunction include retention of sulfobromophthalein and change in serum glutamic-pyruvic transaminase (SGPT) activity following injection[63] or inhalation of methyl chloroform.[62] The doses of methyl chloroform required to cause death and a significant SGPT elevation in 50% of mice within 24 hr are as follows:

Route of administration	24-hr LD_{50}	SGPT activity ED_{50}	SGPT activity Potency ratio LD_{50}/ED_{50}
Intraperitoneal[62]	4.70 g/kg	2.91 g/kg	1.62
Intraperitoneal[63]	4.94 g/kg	3.34 g/kg	1.50
Inhalation[62]	74.25 mg/l for 595 min	74.25 mg/l for $\geq$ 595 min	$\leq$1.00

MacEwen et al.[86] exposed mice continuously for 100 days. The concentration of 1.34 mg/l had no effect on the liver whereas 5.42 mg/l caused an increase in liver weight and elevation of liver triglycerides. The minimal concentration tolerated continuously is between the two levels tested.

In mice, the inhalation for 1 hr of 16.5 mg/l of methyl chloroform caused stimulation of the oxidative activity of microsomal enzymes in the liver, manifested by shortening of sleeping time induced by hexobarbital.[87] These results following inhalation are opposite to those described above for intraperitoneal injection and questions the validity of using sleeping time to indicate hepatic enzyme activity.

The hepatotoxicity has been investigated in rats following inhalation of methyl chloroform (13.75 to 16.5 mg/l) for 24 hr.[88] It decreased the duration of action of hexobarbital, meprobamate and zoxazolamine. There was an in vitro increase in the metabolism of these compounds by hepatic microsomal enzymes under the influence of methyl chloroform. The carbon monoxide-binding pigment (cytochrome p-450 reduced) and nicotinamide adenine dinucleotide phosphate cytochrome reductase activity of the hepatic microsomal fraction were increased. Pretreatment of rats with cycloheximide or actinomycin D prevented the decrease in the hexobarbital narcosis and the increase in hepatic drug metabolism induced by methyl chloroform. It is noteworthy that, after 24 hr of methyl chloroform inhalation, its concentration in the liver was markedly greater than that in the blood.

The inhalation of methyl chloroform in a concentration of 54.28 mg/l for 4 or 6 hr has no effect on liver function of rats fed with ethanol, although there was potentiation of hepatotoxicity of carbon tetrachloride by the alcohol.[89] Pretreatment of rats with phenobarbital sensitized the liver to the hepatotoxicity of carbon tetrachloride but not to methyl chloroform injected intraperitoneally.[90] The intraperitoneal injection of methyl chloroform (1 ml/kg) did not influence liver function in normal rats but was hepatotoxic in alloxan-induced diabetic animals.[91] In the perfused rat liver, methyl chloroform did not influence hepatic blood flow or the morphology of the hepatic parenchymal cells to the same extent as carbon tetrachloride.[92] However, there was an inhibition of respiration of liver mitochondrial preparation following the addition of methyl chloroform.[93]

Oral administration of 1.65 g/kg of methyl chloroform in liquid paraffin to rats for 7 days caused an increase in both microsomal and cell sap protein concentrations in the liver.[94] Intraperitoneal injection of methyl chloroform (3.74 g/kg, i.e., 75% of the LD_{50}) in male rats produced no significant effect on the hepatic triglyceride level, nor was any decrease in hepatic glucose-6-phosphatase activity detected.[64] Disturbances in liver function have been reported in dogs[65] and in rabbits.[95] In all of these investigations, liver function was readily influenced by the administration of methyl chloroform.

Of the laboratory animals investigated, the guinea pig appears most prone to liver injury. While Adams et al.[60] reported no organic injury after 3 months of repeated daily exposure to 8.20 mg/l for 7 hr per day, Torkelson et al.[36] reported the presence of slight lung and liver pathology in guinea pigs exposed repeatedly to 5.5 mg/l for 1.2 hr per day or 11 mg/l for 0.5 hr per day for 3 months. Methyl chloroform produced in mice an enlargement of hepatocytes with cellular infiltration and vacuolation and with slight necrosis only in the lethal range.[63]

Comparison of the hepatotoxic action of the different chlorinated hydrocarbons shows that this increases within each series with the number of chlorine atoms in each molecule. The assumption of some investigators such as Lucas,[96] that the hepatotoxic action is due to liberation of hydrobromic or hydrochloric acid, was not shared by Barrett et al.[97] It appears questionable whether the hepatotoxic action of chlorinated hydrocarbons should be affiliated with the liberation of hydrochloric acid or the formation of phosgene, but it is more likely that it is produced by the molecule *in toto* and it is probably linked to the fat solubility of the solvents.[72]

5. Nephrotoxicity

The renal effects of methyl chloroform have been less extensively studied. This solvent produces definite disturbances of renal function in mice[63,98] and in dogs[65] as shown by phenolsulfonphthalein, glucose and protein excretion data. The kidneys are less vulnerable than the liver to toxic properties of methyl chloroform. Little or no microscopic changes were observed.[63]

6. Dermatotoxicity

Skin contact experiments in rabbits showed only a slight reddening and scaling. Healing occurred promptly when application ceased. Splashing of methyl chloroform into the eyes of rabbits caused slight conjunctival irritation but no corneal damage.[36,99] Absorption of methyl chloroform through the human skin was studied by Stewart and Dodd.[100] They found that the amount of solvent penetrating the skin was related to the area of skin exposed, the method of application, and the duration of exposure. Methyl chloroform absorbed percutaneously is rapidly excreted by the lung.

7. Embryo- and Fetotoxicity

Pregnant mice and rats inhaling methyl chloroform (4.71 mg/l) during gestation showed no elevation of carboxyhemoglobin concentrations nor any significant adverse effects on embryonal and fetal development.[101] Schwetz et al.[102] exposed pregnant rats and mice to 4.75 mg/l for 7 hr daily on days 6 to 15 of gestation. There were no signs of maternal, embryonal, or fetal toxicity.

H. Clinical Toxicology

The diagnosis of exposure to methyl chloroform depends on the detection of the compound in the expired air, blood, or tissue of the individual. If the exposure has been significant, the compound will be present in the expired breath in sufficient concentration to allow specific identification by simple infrared spectrographic techniques in the immediate postexposure period.[41,103,104]

Expired air for infrared analysis may be collected in 6-liter volume Saran® bags. The bagged air is introduced into a long path-length gas cell and the absorbance at 9.2 μm is measured to determine the concentration. The exponential elimination of the compound in the expired air may be followed for a prolonged period of time with the use of the electron captive indicator. Following inhalation of the vapor, the blood concentration decreases exponentially and may not be detected by infrared methods, unless the vapor concentration has approached anesthetic levels.

The treatment of methyl chloroform intoxication includes prompt supportive measures to combat the effects of central nervous system depression.[104] Oxygen with carbon dioxide should be administered. Breathing should be assisted if the respiratory center fails to respond to the carbon dioxide stimulation.

Severe hypotension may be induced by a combination of central nervous system depression and myocardial anoxia secondary to poor oxygen uptake. Unless the situation is desperate, sympathomimetic drugs *must not* be used to combat this hypotension because of the danger of inducing ventricular fibrillation.

I. Threshold Limit Values (TLV) for Methyl Chloroform

The TLV for methyl chloroform, which indicates the safe level, for most workers, of the solvent vapor if it is inhaled over a period of 8 hr daily, 5 days/week which was adopted at the 25th (1963) Annual Meeting of the American Conference of Governmental Industrial Hygienists,[105] was 350 ppm, i.e., 1.93 mg/l.

The following year (1964) the American Industrial Hygiene Association Toxicology Committee[106] evaluated the Emergency Exposure Limits (EEL) for methyl chloroform, which are peak values that should not be exceeded. They are as follows:

Concentration		
ppm	mg/l	Emergency exposure limits min
2500	3.5	5
2000	10.8	15
2000	10.8	30
1000	5.4	60

The TLV of 350 ppm is still in force in the United States at the present time[107,108] and also in France.[109]

It is important to mention that other TLVs have been recommended. Patty[110] has suggested 500 ppm TLV, but the level of 350 ppm is still in force because of complaints due to odor and mild irritation during exposure to low concentrations. Most unacclimated persons can detect 100 ppm (0.55 mg/l); 500 ppm (2.75 mg/l) is definitely noticeable; 1500 ppm (8.25) mg/l) is objectionable; and at 2000 ppm (11 mg/l) the odor is very strong but tolerable. Kay[111] has reviewed the ventilatory requirements in work environment to reduce the concentration of methyl chloroform. It is the opinion of the authors that there has been no compelling observation to alter the current TLV of 350 ppm (1.93 mg/l).

J. Conversion Table for Concentrations of Methyl Chloroform

This review of the literature concludes with a table that is useful in converting concentrations of methyl chloroform on a volume/volume basis to weight/volume. The following relationships are based on the following constants for methyl chloroform: molecular weight of 133.42; volume occupied by one molecular weight at 26.5°C = 24.58 liters; therefore, 5.42857 g of methyl chloroform occupies 1 liter of gas.

mg/l	ppm	v/v%
0.00543	1	–
0.0543	10	0.001
0.1	18.4	–
0.543	100	0.01
1.0	184	–
5.4286	1000	0.1
10	1842	–
54.2857	10,000	1.0
100	18,421	–
271.4285	50,000	5.0
500	92,105	–
542.857	100,000	10.0

Chapter 3

ACUTE INHALATIONAL, ORAL, AND INTRAPERITONEAL TOXICITY OF METHYL CHLOROFORM IN MICE

Studies on experimental animals have been the basis for most of our understanding of the effects of methyl chloroform (1,1,1-trichloroethane) on man. Acute toxicity following oral,[36,69] intraperitoneal,[62-65] or inhalational[60,62] administration of methyl chloroform has been studied in different experimental animals. However, a comparative investigation of methyl chloroform toxicity following its administration to a single animal species via different routes has not yet been reported.

Although central nervous system depression is described as the usual mechanism for methyl chloroform intoxication,[39] this compound has also been shown to have arrhythmogenic properties.[7,10,75-82] The resultant cardiac malfunction is thought to be due either to spontaneous[76] or adrenergic-mediated arrhythmias,[75,82] or both.[7]

In view of the growing commercial importance of this compound and the increased frequency of reported intoxication, it was decided to investigate the role played by the cardiovascular effects in acute intoxication, especially because the toxic properties of this compound cannot be adequately explained as a manifestation of generalized central nervous system depression. For this purpose, the acute lethal $dose_{50}$ (LD_{50}) was determined by the oral and intraperitoneal route, and the lethal $concentration_{50}$ (LC_{50}) by inhalation. Toxicity, furthermore, was correlated with the cardiac effects as reflected by changes in the electrocardiogram.

A. Methods

Experiments were performed on unanesthetized male mice of the Swiss-Webster strain, with weights ranging from 20 to 25 g, divided into three series according to the route of administration of methyl chloroform (stabilized 1,1,1-trichloroethane, Vythene Dupont-TCE):

Series I – Received methyl chloroform by the oral route through a gastric tube. It consisted of five groups with 10 to 15 mice in each group, to which different doses of methyl chloroform were administered (6.7, 9.4, 11.4, 13.4, and 16.7 g/kg). Electrocardiographic changes (lead II) were monitored at repeated intervals up to a period of 72 hr following the administration of methyl chloroform in a sample of 4 mice from Groups 2, 4, and 5, selected at random.

Series II – Received methyl chloroform intraperitoneally. It consisted of five groups with 15 mice in each group, to which different doses of methyl chloroform were administered (0.7, 1.3, 1.7, 2.0, and 2.7 g/kg). Electrocardiographic changes (lead II) were monitored at repeated intervals up to a period of 24 hr in a representative sample of 5 mice, selected at random from groups 1, 2, and 5.

Series III – Received methyl chloroform by inhalation for a period of 5 min. This series consisted of five groups with 10 mice in each group, which were exposed to 0.3, 0.6, 0.8, 1.1, and 1.4 g/l concentrations of methyl chloroform. Various concentrations were prepared by volatilizing, in a known volume of air, a volume of liquid methyl chloroform calculated to give the desired volume of methyl chloroform vapors at standard pressure and 20°C. The process of volatilization was aided by warming the methyl chloroform container. Gaseous concentrations exceeding 0.8 g/l methyl chloroform were balanced with an appropriate volume of oxygen.

In this series, electrocardiographic changes (lead II) were monitored for all animals during the period of exposure and at repeated intervals up to a period of 24 hr after exposure. The animals were individually exposed to the gas in a chamber constructed to allow only nasal exposure to the appropriate gaseous mixture.

Calculation of LD_{50} – Survival was observed in Series I and II for a period of 24 hr and of 48 hr following administration of methyl chloroform. In Series III, however, survival was observed up to a period of 24 hr. The LD_{50} was calculated according to the probit method.[92] Data were analyzed by paired comparisons, the criterion for significance being P less than 0.01.

B. Results

Oral administration – This series consisted of five groups of mice to which varying doses of methyl chloroform were administered orally. Mice selected from Groups 2, 4, and 5 were studied for the electrocardiographic changes, for a period up to 72 hr. Tables 3.1 and 3.2 summarize the data of this series.

Lethal dose$_{50}$ determination: Table 3.1 shows that methyl chloroform in a dose of 6.7 g/kg was not lethal. Doses of 9.4 g/kg, 11.4 g/kg, 13.4 g/kg, and 16.7 g/kg were associated with death rates of 20%, 50%, 60%, and 80% following 24 hr, and of 30%, 60%, 80%, and 100% following 48 hr, respectively. The calculated LD$_{50}$ (95% fiducial limits) for 24 hr were 14.08 ± 2.05 g/kg, and at 48 hr 10.96 ± 1.74 g/kg.

Electrocardiographic changes: Table 3.2 shows that the electrocardiographic changes became more marked with progressively increasing doses of methyl chloroform. Thus, Group 2 (9.4 g/kg) showed a decrease in heart rate averaging 68 ± 17.3 beats/min (±SE diff) and an increase in PR interval averaging 7.8 ± 1.7 msec (±SE diff) 2 hr following the administration of methyl chloroform. Group 4 (13.4 g/kg) showed a more marked decrease in heart rate, starting 1 hr after administration and averaging 185 ± 10.4 beats/min (±SE diff) below control values, and progressively decreasing to an average of 412 ± 13.6 beats/min (±SE diff) 48 hr after administration. The PR interval increased by an average of 21.0 ± 2.0 msec (±SE diff) above control values 1 hr after administration and then continued increasing to an average of 48 ± 1.9 msec (±SE diff) above control values, 48 hr after administration. The QRS potential showed a decrease of 0.05 ± 0 mV and 0.08 ± 0.02 mV (±SE diff) below control average 1 hr and 2 hr after administration, respectively. In Group 5 (16.7 g/kg) the heart rate was decreased by 238 ± 66.7 beats/min as compared with the control average 1 hr after administration, and reached a level of 485 ± 15.0 beats/min below control average 24 hr after administration. The PR interval was prolonged by 36.0 ± 4.0 msec above control average 2 hr after administration, reaching a level of 57 ± 2.0 msec above control average 24 hr after administration. The potential of the QRS complex was decreased 2 hr after administration to a level of 0.08 ± 0.02 mV below control average, reaching a level of 0.25 ± 0.05 mV (±SE diff) below control average 24 hr following administration.

Intraperitoneal injection – Animals in this series, which consisted of five groups, received methyl chloroform intraperitoneally. Mice were randomly selected out of Groups 1, 2, and 5, and were studied for the electrocardiographic changes up to a period of 24 hr. Results are summarized in Tables 3.3 and 3.4.

Lethal dose$_{50}$ determination: Table 3.3 shows an increase in percentage mortality with an increase in dose, reaching 80% with a dose of 2.67 g/kg after 48 hr. The calculated LD$_{50}$ (95% fiducial limits) is 2.34 ± 0.32 g/kg and 1.67 ± 0.29 g/kg at 24 and 48 hr after administration, respectively.

Electrocardiographic changes: Changes in the electrocardiographic measurements were more significant as the dose increased (Table 3.4). Group 1, which received 0.67 g/kg, showed an average decrease in the heart rate of 13 ± 3.0 beats/min, 10 min after injection, that became more manifest as time elapsed, reaching an average decrease of 44 ± 10.8 beats/min 90 min later. Prolongation of the PR interval, averaging 3.0 ± 0.45 msec, was observed 30 min after the injection of methyl chloroform. The significant prolongation of the PR interval continued for 90 min. A

TABLE 3.1
The Effect of Various Doses of Methyl Chloroform, Administered Orally, on the Survival of Mice

Groups	Dose g/kg	Number of mice	Mice dead after 24 hr	Mice dead after 48 hr	LD$_{50}$ (g/kg) 24 hr	LD$_{50}$ (g/kg) 48 hr
1	6.68	15	0			0
2	9.35	10	2	3	14.08	10.96
					±	±
3	11.36	10	5	6	2.05	1.74
4	13.36	10	6			8
5	16.70	10	8			10

TABLE 3.2

Electrocardiographic Changes in Representative Animals of Groups 2, 4 and 5 of Series I Following Oral Administration of Methyl Chloroform*

Procedure	Parameter	C	30 m	1 hr	2 hr	4 hr	6 hr	12 hr	24 hr	48 hr	72 hr
Group 2 (9.4 g/ kg)	HR (beats/ min)	696 ± 7.5	691 ± 6.6 −5 ± 2.0 NS	661 ± 15.6 −35 ± 11.9 NS	628 ± 21.9 −68 ± 17.3 0.01	593 ± 45.5 −104 ± 40.8 NS	535 ± 59.8 −161 ± 54.0 NS	504 ± 69 −193 ± 63.7 NS	533 ± 78.3 −164 ± 74.2 NS	538 ± 102 −159 ± 98.9 NS	677 ± 17.6 −22 ± 14.2 NS
	PR (msec)	26.5 ± 0.5	26.8 ± 0.5 +0.3 ± 0.3 NS	31.0 ± 1.3 +4.5 ± 1.3 NS	34.3 ± 1.6 +7.8 ± 1.7 0.01	39.0 ± 5.0 +12.5 ± 5.1 NS	43.8 ± 6.6 +17.3 ± 6.6 NS	47.8 ± 7.5 +21.3 ± 7.5 NS	45.0 ± 8.7 +18.5 ± 8.7 NS	45.5 ± 13.4 +19.0 ± 13.5 NS	29.0 ± 1.7 +2.3 ± 1.9 NS
	QRS (mV)	0.55 ± 0.02	0.55 ± 0.03 0 ± 0 NS	0.55 ± 0.03 0 ± 0 NS	0.52 ± 0.03 −0.03 ± 0.01 NS	0.51 ± 0.04 −0.04 ± 0.02 NS	0.50 ± 0.05 −0.05 ± 0.03 NS	0.46 ± 0.06 −0.09 ± 0.04 NS	0.45 ± 0.06 −0.09 ± 0.04 NS	0.42 ± 0.08 −0.12 ± 0.06 NS	8.53 ± 0.04 −0.03 ± 0.02 NS
Group 4 (13.4 g/ kg)	HR (beats/ min)	698 ± 10.9	692 ± 6.0 −7 ± 8.8 NS	513 ± 20.3 −185 ± 10.4 0.001	400 ± 11.6 −298 ± 21.2 0.001	387 ± 13.3 −312 ± 24.2 0.001	378 ± 14.2 −320 ± 25.2 0.001	333 ± 8.8 −365 ± 16.1 0.001	320 ± 5.8 −378 ± 7.3 0.001	287 ± 13.3 −412 ± 13.6 0.001	200
	PR (msec)	27.0 ± 0.6	27.3 ± 0.3 +0.3 ± 0.3 NS	48.0 ± 2.1 +21 ± 2.0 0.001	60.0 ± 1.2 +33 ± 1.7 0.001	63.3 ± 0.7 +36 ± 1.2 0.001	64.7 ± 0.7 +37 ± 0.9 0.001	67.3 ± 0.7 +40 ± 0.9 0.001	70.0 ± 1.2 +43 ± 0.1 0.001	75.3 ± 2.4 +48 ± 1.9 0.001	86
	QRS (mV)	0.50 ± 0.02	0.50 ± 0.03 0 ± 0 NS	0.45 ± 0.03 −0.05 ± 0 0.001	0.42 ± 0.02 −0.08 ± 0.02 0.01	0.40 ± 0.0 −0.10 ± 0.03 NS	0.38 ± 0.02 −0.12 ± 0.03 NS	0.35 ± 0.03 −0.15 ± 0.05 NS	0.33 ± 0.03 −0.017 ± 0.06 NS	0.30 ± 0.03 −0.20 ± 0.05 NS	0.20
Group 5 (16.7 g/ kg)	HR (beats/ min)	708 ± 6.0	693 ± 8.8 −15 ± 13.2 NS	470 ± 65.6 −238 ± 66.7 0.02	417 ± 49.8 −292 ± 49.7 0.01	303 ± 12.0 −415 ± 20.2 0.001	300 ± 10.0 −408 ± 10.1 0.001	257 ± 28.5 −452 ± 26.8 0.001	225 ± 25.0 −485 ± 15.0 0.001		
	PR (msec)	26.0 ± 0.6	28.3 ± 0.9 +2 ± 1.3 NS	53.0 ± 9.6 +27 ± 9.5 NS	62.0 ± 4.2 +36 ± 4.0 0.001	72.7 ± 3.7 +47 ± 3.7 0.001	72.7 ± 3.7 +47 ± 3.7 0.001	79.0 ± 2.6 +53 ± 2.5 0.001	83.0 ± 3.0 +57 ± 2.0 0.001		
	QRS (mV)	0.48 ± 0.02	0.50 ± 0.03 +0.02 ± 0.02 NS	0.45 ± 0.03 −0.03 ± 0.02 NS	0.40 ± 0 −0.08 ± 0.02 0.01	0.35 ± 0.03 −0.13 ± 0.03 NS	0.33 ± 0.03 −0.15 ± 0.03 0.01	0.27 ± 0.03 −0.22 ± 0.04 0.01	0.22 ± 0.02 −0.25 ± 0.05 0.01		

*Each group of numbers represents, in order, mean ± SEM, mean difference ± SE of difference, and the P value. C = control, HR = heart rate, PR = PR interval, and QRS = QRS potential.

TABLE 3.3

The Effect of Various Doses of Methyl Chloroform, Injected Intraperitoneally, on the Survival of Mice

Groups	Dose g/kg	Number of mice	Mice dead after 24 hr	Mice dead after 48 hr	LD$_{50}$ (g/kg) 24 hr	LD$_{50}$ (g/kg) 48 hr
1	0.67	15	1	2		
2	1.34	15	1	3		
3	1.67	15	2	7	2.34 ± 0.32	1.67 ± 0.29
4	2.01	15	5	11		
5	2.67	15	9	12		

considerable decrease in heart rate, averaging 98 ± 20.6 beats/min, started 10 min after the administration of 1.34 g/kg of methyl chloroform to mice of Group 2. This decrease progressed and reached a value of 226 ± 35.4 beats/min after 6 hr. Significant prolongation of the PR interval started 5 min after methyl chloroform administration and lasted for 6 hr.

In Group 5 (2.67 g/kg), however, a significant decrease in the heart rate, averaging 47 ± 8.9, was observed 5 min after injection and gradually intensified for a period of 6 hr when all the animals died. A maximum decrease in heart rate of 439 ± 51.6 beats/min and a maximum increase in PR interval of 53.4 ± 4.5 msec were observed after 2 hr. It is only in this group that a significant decrease in the QRS potential, averaging 0.11 ± 0.02 mV, was exhibited 10 min after administration of methyl chloroform. This decrease became gradually more apparent until it reached 0.33 ± 0.03 mV after 6 hr, culminating in the death of all the mice.

Inhalation – This series consisted of five groups of mice which were exposed to different concentrations of methyl chloroform via inhalation route. All the animals were studied for electrocardiographic changes up to a period of 24 hr. The results are summarized in Tables 3.5 and 3.6.

Lethal dose$_{50}$ determination: Results depicted in Table 3.5 show that inhalation of a concentration of 0.28 g/l of methyl chloroform caused 10% death; this percentage is gradually increased by raising the vapor concentration of methyl chloroform. A 1.39 g/l concentration caused 90% death. The median lethal concentration for 24 hr after exposure to methyl chloroform vapors was 0.81 ± 0.14 g/l as calculated by the probit method.

Electrocardiographic changes: In Group 1 (0.28 g/l) the only significant change observed was a decrease in heart rate which started 2 min after the beginning of exposure, averaging 84 ± 21.0 beats/min, and persisted during the period of exposure, reaching an average decrease of 191 ± 44.5 beats/min below the control at the end of exposure. In Group 2 (0.56 g/l), however, a significant decrease in heart rate, averaging 167 ± 54.2 beats/min, was observed only at the fifth minute of exposure. Nevertheless, a significant prolongation in PR interval, averaging 5.9 ± 2.1 msec, was observed 1 min after the start of exposure and continued throughout the exposure, where it reached a maximum duration of 14.3 ± 8.7 msec longer than the control average.

The Group 3 mice, which breathed 0.83 g/l of methyl chloroform, showed a decrease in heart rate that started at the beginning of exposure and gradually intensified, reaching an average value of 187 ± 82.6 beats/min at the end of exposure. However, a significant increase in heart rate was observed 20 to 30 min after exposure. A prolongation in the PR interval was observed reaching a maximum of 28.4 ± 12.24 msec at the third minute of exposure and the effect was maintained throughout the exposure period. A decrease in the QRS potential was observed during exposure time, reaching a value of 0.13 ± 0.04 mV below the control value at the second minute of exposure.

Inhalation of 1.11 g/l of methyl chloroform brought about in animals of Group 4 a significant decrease in heart rate averaging 233 ± 80.1 beats/min below the control at the fifth minute of exposure. A significant prolongation of the PR interval was observed during the exposure time, reaching a maximum of 20.8 ± 8.74 msec at the fourth minute of exposure. A significant decrease

TABLE 3.4

Electrocardiographic Changes in Representative Animals of Groups 1, 2 and 5 of Series II after Intraperitoneal Injection of Methyl Chloroform*

Procedure	Parameter	C	5 m	10 m	20 m	30 m	60 m	90 m	120 m	6 hr	24 hr
Group 1	HR	701 ± 4.0	700 ± 3.5	688 ± 4.9	663 ± 11.4	648 ± 10.7	654 ± 9.3	657 ± 8.3	664 ± 7.5	678 ± 4.9	692 ± 3.7
(0.67 g/	(beats/		−1.0 ± 2.4	−13 ± 3.0	−38 ± 12.8	−53 ± 11.1	−47 ± 11.1	−44 ± 10.8	−37 ± 11.1	−23 ± 7.7	−9 ± 5.1
kg)	min)		NS	0.01	0.02	0.01	0.01	0.01	0.02	0.02	NS
	PR	25.2 ± 0.3	24.8 ± 0.3	25.4 ± 0.2	27.2 ± 0.4	28.2 ± 0.5	27.2 ± 0.4	27.2 ± 0.4	26.8 ± 0.4	26.2 ± 0.4	25.4 ± 0.4
	(msec)		−0.4 ± 0.24	+0.2 ± 0.37	+2.0 ± 0.45	+3.0 ± 0.45	+2.0 ± 0.55	+2.0 ± 0.55	+1.6 ± 0.68	+1.0 ± 0.71	+0.2 ± 0.37
			NS	NS	0.01	0.001	0.01	0.01	NS	NS	NS
	QRS	0.61 ± 0.02	0.61 ± 0.02	0.61 ± 0.01	0.56 ± 0.02	0.53 ± 0.02	0.54 ± 0.02	0.54 ± 0.02	0.56 ± 0.02	0.57 ± 0.02	0.57 ± 0.02
	(mV)		0 ± 0	0 ± 0.02	−0.05 ± 0.03	−0.08 ± 0.03	−0.07 ± 0.03	−0.07 ± 0.03	−0.05 ± 0.02	−0.04 ± 0.02	0.03 ± 0.02
			NS	NS	NS	NS	NS	NS	NS	NS	NS
Group 2	HR	713 ± 4.8	700 ± 9.1	615 ± 23.3	570 ± 26.8	518 ± 33.3	508 ± 32.8	486 ± 27.5	485 ± 28.7	487 ± 37	540 ± 106.9
(1.34 g/	(beats/		−13 ± 6.3	−98 ± 20.6	−143 ± 25.6	−195 ± 31.8	−205 ± 31.8	−226 ± 26.3	−228 ± 27.5	−226 ± 35.4	−173 ± 105.9
kg)	min)		NS	0.01	0.01	0.001	0.001	0.001	0.001	0.001	NS
	PR	24.3 ± 0.3	26.8 ± 0.5	34.0 ± 2.3	38.5 ± 3.3	44.5 ± 4.0	46.3 ± 4.2	50.8 ± 2.5	51.0 ± 2.4	51.0 ± 4.4	47.3 ± 14.3
	(msec)		+2.5 ± 0.5	+9.8 ± 2.3	+14.3 ± 3.4	+20.3 ± 4.0	+22.0 ± 4.2	+26.5 ± 2.5	+26.8 ± 2.4	+26.8 ± 4.5	+23.0 ± 14.3
			0.01	0.02	0.01	0.01	0.01	0.001	0.001	0.001	NS
	QRS	0.61 ± 0.01	0.61 ± 0.01	0.60 ± 0.0	0.56 ± 0.02	0.54 ± 0.02	0.49 ± 0.04	0.49 ± 0.04	0.47 ± 0.04	0.45 ± 0.05	0.45 ± 0.07
	(mV)		0 ± 0	−0.01 ± 0.01	−0.08 ± 0.01	−0.08 ± 0.03	−0.13 ± 0.05	−0.13 ± 0.05	−0.14 ± 0.05	−0.16 ± 0.06	−0.16 ± 0.07
			NS	NS	0.01	NS	NS	NS	NS	NS	NS
Group 5	HR	711 ± 11.0	664 ± 14.4	591 ± 8.7	463 ± 11.2	421 ± 11.8	394 ± 8.1	298 ± 50.7	272 ± 49.2	290 ± 4.1	
(2.67 g/	(beats/		−47 ± 8.9	−120 ± 7.7	−248 ± 10.5	−290 ± 14.9	−317 ± 13.0	−413 ± 52.6	−439 ± 51.6	−419 ± 14.2	
kg)	min)		0.001	0.001	0.001	0.001	0.001	0.001	0.001	0.001	
	PR	27.8 ± 0.9	36.2 ± 1.4	46.8 ± 1.5	57.4 ± 1.1	65.0 ± 1.5	69.6 ± 1.5	77.2 ± 3.7	81.2 ± 4.3	80.5 ± 0.5	
	(msec)		+8.4 ± 0.75	+19.0 ± 1.34	+29.6 ± 1.29	+37.2 ± 1.46	+41.8 ± 2.06	+49.4 ± 3.92	+53.4 ± 4.51	+52.5 ± 1.5	
			0.001	0.001	0.001	0.001	0.001	0.001	0.001	0.001	
	QRS	0.65 ± 0.02	0.60 ± 0.03	0.54 ± 0.02	0.49 ± 0.03	0.45 ± 0.02	0.45 ± 0.02	0.39 ± 0.02	0.35 ± 0.03	0.31 ± 0.01	
	(mV)		−0.05 ± 0.02	−0.11 ± 0.02	−0.16 ± 0.03	−0.20 ± 0.03	−0.20 ± 0.03	−0.26 ± 0.03	−0.30 ± 0.04	−0.33 ± 0.03	
			NS	0.01	0.001	0.001	0.001	0.001	0.001	0.001	

*Each group of numbers represents, in order, mean ± SEM, mean difference ± SE of difference and the P value. C = control, HR = heart rate, PR = PR interval, and QRS = QRS potential.

TABLE 3.5

The Effect of Various Concentrations of Methyl Chloroform Administered by Inhalation on the Survival of Mice

Groups	Calculated vapor concentration % v/v	g/l	Number of mice	Mice dead after 24 hr	LC_{50} (g/l) 24 hr
1	5	0.28	10	1	
2	10	0.56	10	3	
3	15	0.83	10	5	0.81 ± 0.14
4	20	1.11	10	6	
5	25	1.39	10	9	

in QRS potential of 0.12 ± 0.03 mV was reached at the third minute of exposure.

In the fifth group, which inhaled 1.39 g/l of methyl chloroform, a significant decrease was observed in all electrocardiographic parameters studied. Thus, a decrease in heart rate of 162 ± 38.7 beats/min below control value at the beginning of exposure was gradually aggravated, reaching an average decrease of 258 ± 38.2 beats/min at the end of exposure. Significant prolongation of the PR interval was observed, starting from the second minute of exposure and reaching a maximum of 39.0 ± 15.0 msec. The QRS potential was significantly decreased throughout the exposure period. In the first minute of exposure a decrease of 0.13 ± 0.05 mV below the control value was observed, which gradually intensified to reach a peak decrease of 0.29 ± 0.09 mV in the fifth minute. Meanwhile, a widening of the QRS complex by 1.6 ± 0.5 msec above the control value, which gradually increased to reach an average of 7.0 ± 1.5 msec, was noticed.

C. Discussion

1. Lethal Dose and Lethal Concentration$_{50}$

The results of this study show the comparative amounts of methyl chloroform that are lethal to mice when ingested orally, injected intraperitoneally, or inhaled. The single oral dose that caused 50% mortality of mice after 24 hr was 14.08 ± 2.05 g/kg. This value is in agreement with that reported by Torkelson et al.[36] who found that 11.24 g/kg and 9.7 g/kg of uninhibited and inhibited methyl chloroform, respectively, were lethal to 50% of female mice. However, the slight difference between the LD_{50} in the present study and that determined by Torkelson et al.[36] might be due to differences in sex. Furthermore, Rowe et al.[69] found that 10.3 g/kg of a solvent mixture containing 74% methyl chloroform killed 50% of female mice.

In the present study, the LD_{50} calculated at the end of 24 hr after intraperitoneal administration is 2.34 ± 0.32 g/kg. However, Klaassen and Plaa,[63] using male mice, and Gehring,[62] using female mice, estimated the LD_{50} for the same solvent after 24 hr of intraperitoneal administration as 37 mM/kg (i.e., 4.94 g/kg) and 35.2 mM/kg (i.e., 4.69 g/kg), respectively. There is a significant difference ($P < 0.01$) between the LD_{50} value obtained in the present study and those obtained by Gehring[62] and by Klaassen and Plaa.[63]

This discrepancy can readily be explained by the fact that the aforementioned authors administered methyl chloroform dissolved in corn oil, whereas in the present study it was administered undiluted. It is probable that the absorption of methyl chloroform by the peritoneal mucosal surface is delayed by an oily vehicle. Further evidence that this is the case is shown by the work of Takeuchi,[61] who administered pure methyl chloroform to mice and reported an LD_{50} of 2.56 g/kg, a value which is not significantly different from our findings.

The LC_{50} of methyl chloroform administered by inhalation for a period of 5 min was found, in the present study, to be 0.81 ± 0.14 g/l. Other investigators have reported LC_{50} values for methyl chloroform administered by inhalation in both mice[62] and rats[60]. However, comparison with our data is not possible because of differences in duration of exposure.

2. Electrocardiographic Changes

Administration of methyl chloroform by the

TABLE 3.6

Electrocardiographic Changes During (1–5 min) and at Different Intervals Following Inhalation of Methyl Chloroform*

Procedure	Parameter	C	1m	2m	3m	4m	5m	10m	20m	30m	60m	90m	120m	6hr	24hr
Group 1 (0.28 g/l)	HR (beats/min)	609 ± 20.5	609 ± 17.6 0 ± 18.7 NS	525 ± 25.1 −84 ± 21.0 0.001	497 ± 37.6 −122 ± 37.2 0.01	437 ± 45.3 −182 ± 40.9 0.001	428 ± 46.0 −191 ± 44.5 0.001	623 ± 28.5 +4 ± 26.5 NS	600 ± 23.4 −19 ± 19.1 NS	596 ± 15.8 −23 ± 17.7 NS	623 ± 26.0 +4 ± 22.1 NS	623 ± 24.4 +17 ± 19.1 NS	654 ± 20.1 +36 ± 18.9 NS	597 ± 17.6 −22 ± 12.3 NS	558 ± 12.6 −61 ± 22.1 0.02
	PR (msec)	29.7 ± 1.13	32.6 ± 3.40 2.9 ± 2.79 NS	30.8 ± 1.35 1.7 ± 1.30 NS	35.2 ± 2.29 6.1 ± 2.78 NS	47.0 ± 6.59 +17.9 ± 6.74 0.02	49.8 ± 8.97 +20.7 ± 9.04 NS	31.9 ± 1.43 +2.8 ± 1.40 NS	29.8 ± 0.76 +0.7 ± 1.20 NS	28.9 ± 0.48 +0.2 ± 1.19 NS	29.0 ± 1.00 +0.1 ± 1.37 NS	27.3 ± 1.05 −1.8 ± 1.19 NS	26.2 ± 1.15 −2.9 ± 1.80 NS	28.7 ± 1.12 −0.4 ± 0.85 NS	28.9 ± 0.79 −0.2 ± 0.91 NS
	QRS (p) (mV)	0.51 ± 0.03	0.56 ± 0.04 +0.05 ± 0.03 NS	0.54 ± 0.04 +0.03 ± 0.02 NS	0.56 ± 0.03 +0.05 ± 0.03 NS	0.56 ± 0.05 +0.05 ± 0.05 NS	0.55 ± 0.05 +0.04 ± 0.05 NS	0.60 ± 0.06 +0.09 ± 0.04 NS	0.56 ± 0.07 +0.05 ± 0.05 NS	0.56 ± 0.07 +0.05 ± 0.07 NS	0.56 ± 0.06 +0.05 ± 0.06 NS	0.57 ± 0.06 +0.05 ± 0.05 NS	0.52 ± 0.07 +0.01 ± 0.06 NS	0.48 ± 0.04 +0.03 ± 0.05 NS	0.46 ± 0.03 +0.06 ± 0.04 NS
	QRS (d) (msec)	9.3 ± 0.33	10.3 ± 0.94 1.0 ± 0.73 NS	11.2 ± 1.09 1.9 ± 0.96 NS	11.2 ± 1.22 +2.0 ± 1.24 NS	12.1 ± 1.46 +2.9 ± 1.44 NS	12.4 ± 1.62 +3.2 ± 1.65 NS	10.7 ± 0.73 +1.4 ± 0.67 NS	10.4 ± 0.78 +1.2 ± 0.59 NS	10.7 ± 0.80 +1.4 ± 0.60 NS	10.6 ± 0.41 +1.3 ± 0.47 NS	10.0 ± 0.62 +0.8 ± 0.49 NS	9.6 ± 0.5 +0.3 ± 0.53 NS	10.3 ± 0.88 +1.1 ± 0.82 NS	9.3 ± 0.44 +0.1 ± 0.65 NS
Group 2 (0.56 g/l)	HR (beats/min)	575 ± 17.4	588 ± 25.6 +13 ± 28.2 NS	574 ± 44.6 −1 ± 43.6 NS	527 ± 43.4 −48 ± 41.0 NS	474 ± 41.2 −101 ± 41.9 NS	508 ± 83.6 −167 ± 54.2 0.01	550 ± 31.9 −14 ± 26.8 NS	564 ± 24.9 0 ± 29.1 NS	571 ± 19.3 +7 ± 19.6 NS	577 ± 33.1 +13 ± 35.0 NS	579 ± 32.4 +14 ± 36.0 NS	586 ± 35.3 +21 ± 31.4 NS	584 ± 28.2 +20 ± 29.8 NS	560 ± 24.2 −4 ± 17.1 NS
	PR (msec)	26.1 ± 0.43	32.0 ± 2.34 +5.9 ± 2.1 0.02	34.5 ± 3.14 +8.4 ± 2.9 0.01	38.9 ± 3.94 +12.6 ± 3.8 0.01	35.7 ± 2.29 +9.7 ± 2.5 0.01	40.3 ± 8.15 +14.3 ± 8.7 0.01	27.9 ± 0.88 +1.6 ± 0.9 NS	27.9 ± 1.01 +1.6 ± 0.9 NS	31.1 ± 0.77 +4.9 ± 9.7 NS	29.6 ± 1.49 +3.3 ± 1.6 NS	26.7 ± 1.07 +0.3 ± 1.0 NS	28.3 ± 0.89 +2.0 ± 0.9 NS	27.0 ± 1.09 +0.7 ± 1.1 NS	26.9 ± 1.20 +0.6 ± 1.1 NS
	QRS (p) (mV)	0.44 ± 0.03	0.42 ± 0.03 −0.02 ± 0.02 NS	0.42 ± 0.04 −0.0 2± 0.03 NS	0.41 ± 0.03 −0.03 ± 0.02 NS	0.40 ± 0.04 −0.04 ± 0.03 NS	0.39 ± 0.04 −0.05 ± 0.03 NS	0.44 ± 0.04 0 ± 0.03 NS	0.44 ± 0.03 0 ± 0.03 NS	0.46 ± 0.02 +0.02 ± 0.03 NS	0.45 ± 0.03 +0.01 ± 0.03 NS	0.47 ± 0.03 +0.03 ± 0.04 NS	0.42 ± 0.04 −0.02 ± 0.04 NS	0.47 ± 0.03 +0.03 ± 0.03 NS	0.50 ± 0.03 +0.06 ± 0.03 NS
	QRS (d) (msec)	13.7 ± 0.67	12.8 ± 1.00 −0.9 ± 0.97 NS	14.6 ± 1.20 +0.9 ± 1.3 NS	14.7 ± 1.30 +1.0 ± 1.3 NS	14.0 ± 1.14 +0.3 ± 1.1 NS	16.3 ± 1.41 +2.6 ± 1.6 NS	15.1 ± 1.08 +1.1 ± 0.7 NS	16.9 ± 1.18 +2.9 ± 1.8 NS	15.9 ± 1.24 +1.9 ± 1.2 NS	15.3 ± 1.19 +1.3 ± 1.6 NS	17.4 ± 1.27 +3.4 ± 2.1 NS	16.6 ± 1.27 +2.6 ± 1.7 NS	12.6 ± 1.02 −1.4 ± 1.4 NS	11.9 ± 1.03 −2.1 ± 1.4 NS
Group 3 (0.83 g/l)	HR (beats/min)	599 ± 22.6	561 ± 66.3 −38 ± 68.2 NS	531 ± 69.6 −68 ± 73.9 NS	506 ± 80.3 −93 ± 87.5 NS	464 ± 85.8 −135 ± 91.9 0.01	412 ± 80.1 −187 ± 82.6 0.01	660 ± 20.9 +82 ± 36.3 NS	680 ± 22.8 +102 ± 24.6 0.01	658 ± 24.9 +80 ± 13.0 0.01	606 ± 29.4 +28 ± 25.6 NS	628 ± 40.2 +50 ± 25.5 NS	608 ± 30.1 +30 ± 20.7 NS	630 ± 38.9 +52 ± 20.3 NS	594 ± 33.5 +16 ± 33.3 NS
	PR (msec)	30.4 ± 1.04	43 ± 2.96 +12.5 ± 2.46 0.001	47.6 ± 4.70 +16.5 ± 4.02 0.001	60.0 ± 13.1 +28.4 ± 12.24 0.01	39.4 ± 1.17 +9.2 ± 1.16 0.001	38.2 ± 2.03 +8.0 ± 1.92 0.01	30.4 ± 0.24 +0.2 ± 0.37 NS	30.2 ± 0.37 0 ± 0.32 NS	30.8 ± 0.37 +0.6 ± 0.51 NS	30.8 ± 1.49 +0.8 ± 1.49 NS	30.8 ± 0.58 +0.6 ± 0.6 NS	29.4 ± 0.87 −0.8 ± 0.86 NS	30.4 ± 0.24 +0.2 ± 0.37 NS	31.4 ± 0.98 +1.2 ± 1.07 NS
	QRS (p) (mV)	0.58 ± 0.05	0.47 ± 0.04 −0.12 ± 0.02 0.001	0.47 ± 0.06 −0.13 ± 0.04 0.01	0.47 ± 0.06 −0.11 ± 0.04 0.01	0.48 ± 0.07 −0.10 ± 0.06 NS	0.51 ± 0.09 −0.07 ± 0.07 NS	0.47 ± 0.06 −0.18 ± 0.06 NS	0.44 ± 0.05 −0.20 ± 0.09 NS	0.56 ± 0.04 −0.09 ± 0.11 NS	0.55 ± 0.05 −0.10 ± 0.10 NS	0.52 ± 0.03 −0.12 ± 0.07 NS	0.48 ± 0.04 −0.17 ± 0.10 NS	0.54 ± 0.03 −0.11 ± 0.08 NS	0.46 ± 0.02 −0.19 ± 0.07 NS
	QRS (d) (msec)	11.2 ± 1.09	14.7 ± 1.46 +3.4 ± 1.39 NS	13.8 ± 1.52 +2.4 ± 1.43 NS	15.7 ± 1.61 +4.4 ± 1.81 NS	16.1 ± 1.67 +4.8 ± 1.82 0.02	15.6 ± 1.48 +4.3 ± 1.53 0.02	12.0 ± 1.10 +0.4 ± 1.69 NS	12.8 ± 1.39 +0.4 ± 1.43 NS	12.6 ± 0.60 +0.1 ± 1.74 NS	11.6 ± 0.93 +0.8 ± 1.65 NS	10.4 ± 0.40 −2.0 ± 1.58 NS	10.2 ± 1.02 +2.2 ± 1.59 NS	11.0 ± 0.55 −1.4 ± 1.99 NS	10.8 ± 0.49 −1.6 ± 2.01 NS
Group 4 (1.11 g/l)	HR (beats/min)	644 ± 17.6	569 ± 44.7 −84 ± 42.1 NS	548 ± 54.5 −94 ± 50.2 NS	476 ± 44.5 −140 ± 36.1 0.01	448 ± 60.6 −203 ± 64.3 0.01	414 ± 78.3 −233 ± 80.1 0.01	528 ± 35.4 −110 ± 55.7 0.02	628 ± 30.9 −10 ± 38.1 NS	628 ± 21.4 −10 ± 38.1 NS	645 ± 14.4 −7.5 ± 35.4 NS	638 ± 41.7 0 ± 61.2 NS	613 ± 57.6 −25 ± 69.1 NS	635 ± 18.5 +2.5 ± 40.7 NS	643 ± 33.3 +5 ± 50.1 NS
	PR (msec)	28.1 ± 1.27	37.9 ± 2.26 +11.3 ± 2.26 0.001	34.9 ± 1.85 +7.6 ± 2.70 0.01	35.6 ± 3.56 +8.4 ± 3.95 0.01	48.0 ± 7.91 +20.8 ± 8.74 0.01	36.5 ± 5.19 +8.8 ± 3.1 0.01	33.8 ± 4.97 +6.0 ± 4.92 NS	30.0 ± 2.89 +2.3 ± 3.93 NS	27.5 ± 1.44 +0.3 ± 1.89 NS	31.7 ± 1.67 +4.0 ± 3.06 NS	30.0 ± 2.20 +3.0 ± 1.03 NS	26.7 ± 1.67 −1.0 ± 1.10 NS	30.8 ± 1.49 +3.0 ± 2.38 NS	28.8 ± 3.19 +1.0 ± 4.18 NS
	QRS (p) (mV)	0.45 ± 0.04	0.39 ± 0.05 −0.07 ± 0.03 0.01	0.41 ± 0.04 −0.06 ± 0.03 0.01	0.34 ± 0.05 −0.12 ± 0.03 0.01	0.35 ± 0.06 −0.09 ± 0.05 NS	0.39 ± 0.05 −0.05 ± 0.05 NS	0.39 ± 0.06 −0.08 ± 0.03 NS	0.39 ± 0.07 −0.08 ± 0.04 NS	0.37 ± 0.05 −0.10 ± 0.06 NS	0.39 ± 0.08 −0.08 ± 0.11 NS	0.45 ± 0.10 −0.03 ± 0.03 NS	0.45 ± 0.04 −0.03 ± 0.04 NS	0.46 ± 0.06 −0.02 ± 0.03 NS	0.48 ± 0.09 +0.01 ± 0.04 NS
	QRS (d) (msec)	12.8 ± 1.09	13.9 ± 1.12 +1.3 ± 1.2 NS	13.4 ± 1.11 +1.1 ± 1.14 NS	15.9 ± 1.49 +2.5 ± 1.86 NS	17.5 ± 1.50 +4.3 ± 1.81 NS	16.4 ± 2.23 +2.7 ± 1.74 NS	15.8 ± 1.49 +2.5 ± 0.87 0.02	13.5 ± 1.66 +0.3 ± 3.01 NS	14.0 ± 0.71 +0.8 ± 1.25 NS	13.5 ± 0.65 +0.3 ± 1.25 NS	13.8 ± 1.25 +0.5 ± 1.85 NS	15.5 ± 0.87 +2.3 ± 1.88 NS	14.0 ± 1.41 +0.8 ± 1.11 NS	13.8 ± 1.65 +0.5 ± 1.32 NS

TABLE 3.6 (continued)

Electrocardiographic Changes During (1–5 min) and at Different Intervals Following Inhalation of Methyl Chloroform*

Procedure	Parameter	C	1m	2m	3m	4m	5m	10m	20m	30m	60m	90m	120m	6hr	24hr
Group	HR	586 ± 26.8	424 ± 50.8	374 ± 64.3	314 ± 59.1	414 ± 60.3	363 ± 72.4								
5	(beats/		−162 ± 38.7	−201 ± 42.6	−250 ± 40.7	−202 ± 41.2	−258 ± 38.2								
(1.39			0.001	0.001	0.001	0.01	0.001								
g/l)															
	PR	30.7 ± 0.67	46.8 ± 7.19	58.0 ± 10.12	69.0 ± 14.98	52.3 ± 6.77	56.5 ± 6.5								
	(msec)		+16 ± 7.04	+27 ± 10.0	39.0 ± 15.0	+22.0 ± 6.82	+24 ± 4.0								
			NS	0.02	0.02	0.02	0.02								
	QRS	0.64 ± 0.04	0.51 ± 0.05	0.42 ± 0.06	0.36 ± 0.05	0.38 ± 0.04	0.39 ± 0.11								
	(p)		−0.13 ± 0.05	−0.22 ± 0.07	−0.25 ± 0.03	−0.28 ± 0.04	−0.29 ± 0.09								
	(mV)		0.02	0.01	0.001	0.001	0.001								
	QRS	9.8 ± 0.23	11.4 ± 0.67	13.1 ± 0.93	15.3 ± 1.11	16.0 ± 0.77	16.8 ± 1.37								
	(d)		+1.6 ± 0.5	+3.4 ± 0.9	+5.7 ± 0.9	+6.4 ± 0.8	+7.0 ± 1.5								
	(msec)		0.01	0.01	0.001	0.001	0.01								

*HR = heart rate
PR = PR interval
QRS (p) = QRS potential
and QRS (d) = duration of QRS complex.

oral, intraperitoneal, or pulmonary routes brought about a dose-dependent decrease in heart rate (Figure 3.1). This is in line with findings by Herd et al.[83] However, in the group of mice which inhaled a concentration of methyl chloroform of 0.83 g/l, an increase in heart rate was observed 15 to 25 min after the end of exposure (Table 3.6). In the absence of simultaneous hemodynamic measurements, the explanation of this rebound phenomenon can only be speculative. It is conceivable that the residual concentration of methyl chloroform at this stage following exposure was enough to bring about hypotension and with it reflex sympathoadrenal discharge, thus overshadowing methyl chloroform-induced bradycardia.

The decrease in heart rate was accompanied by an increase of the AV conduction time as reflected by prolongation of the PR interval (Figure 3.2). In addition, the following was observed: a first and second degree AV block, a decrease in intraventricular conduction (Figure 3.3) as well as notched QRS complex, depressed S-T segment, flattened T wave and premature ventricular contractions. This is in agreement with reported studies.[7,45,75] A significant decrease in the QRS potential was observed, especially with high doses, viz., 13.4 and 16.7 g/kg administered orally, 2.67 g/kg intraperitoneally, and 0.83, 1.11 and 1.19 g/l by inhalation (Figure 3.4).

A striking feature is that the decrease in the QRS potential and the widening of the QRS

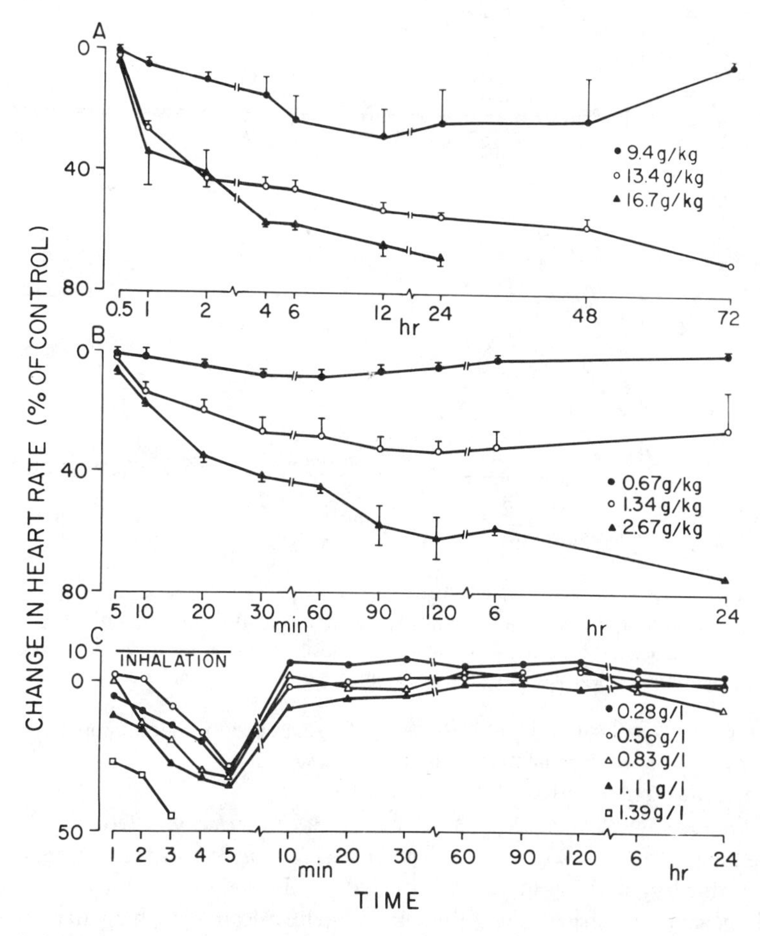

FIGURE 3.1 Effect of different doses of methyl chloroform on heart rate of unanesthetized mice, after oral administration (A), intraperitoneal injection (B), or inhalation (C). A dose-dependent decrease in heart rate is observed. Each point represents the mean of 4 mice (A), 5 mice (B), and 10 mice (C). Vertical bars representing SE of mean were omitted in (C) for simplicity.

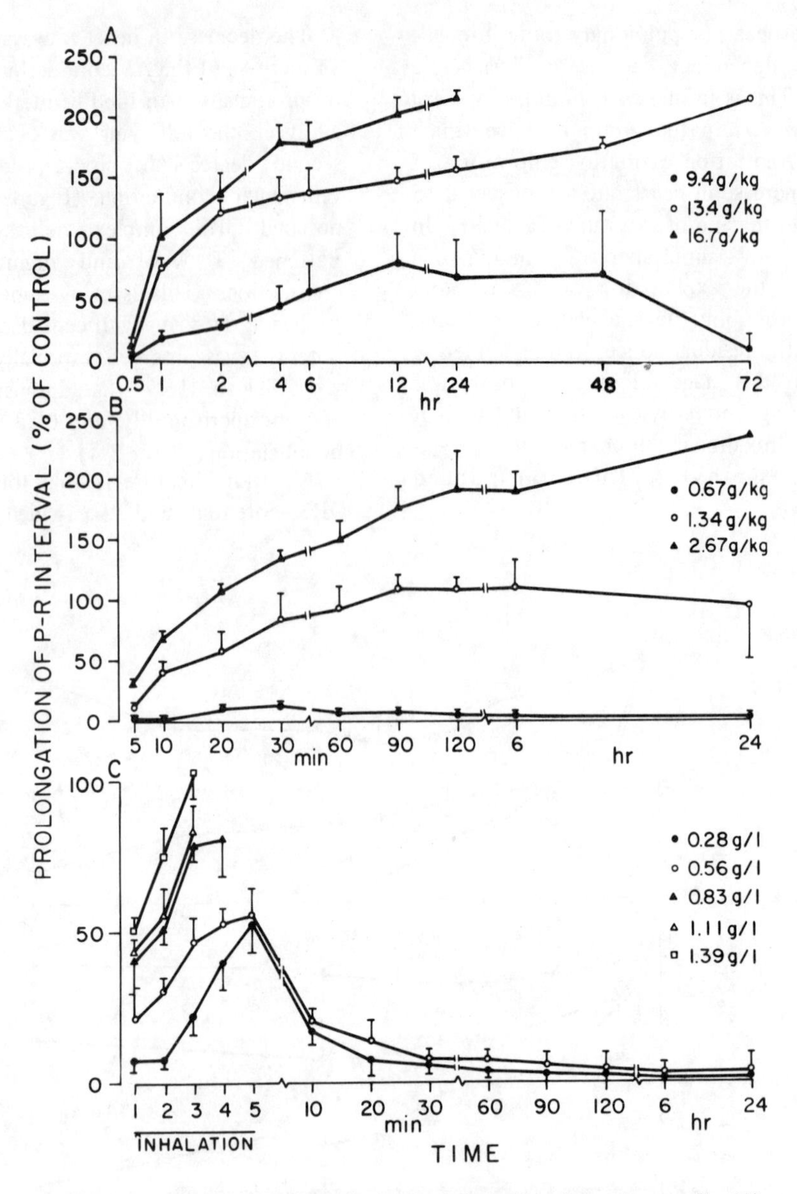

FIGURE 3.2 Effect of various doses of methyl chloroform given orally (A), intraperitoneally (B), or by inhalation (C), on the PR interval. Note the dose-dependent prolongation of PR interval. Number of mice is the same mentioned in legend of Figure 3.1. All groups of mice had similar control values.

complex, unlike changes in heart rate and the PR interval, lasted for 24 hr after inhalation (Figure 3.5). This reflects deleterious effects of methyl chloroform on the heart muscle which is detected exclusively by changes in the QRS complex. This was further confirmed by plotting the percentage mortality and percentage change in different parameters of the electrocardiogram against different doses. Results are depicted in Figure 3.6, which reveals that mortality correlates best with changes in the QRS potential and duration, implying that lethality at high doses may be a consequence of depression of myocardial contracti ity.

3. Correlation Between Lethal Amounts an Electrocardiographic Changes

In the present study, the marked changes of th electrocardiographic parameters brought about b increase in doses were generally accompanied by a increase in mortality rate. This stirred our intere to find the correlation between LD_{50} and LC_5 and the electrocardiographic changes. The calcula ed LC_{50} after inhalation of methyl chlorofor

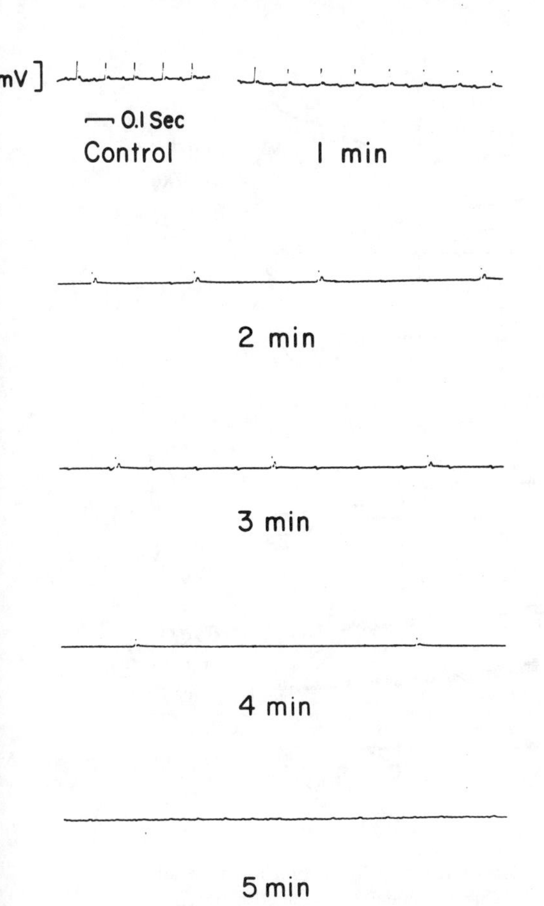

FIGURE 3.3 The effects of breathing 1.4 g/l methyl chloroform for 5 min on the electrocardiogram (lead II) of a mouse. Note prolongation of PR interval and decrease in heart rate after 1 min; A-V block, decrease in QRS potential and widening of QRS complex after 2 to 3 min; complete A-V block after 4 min; and death after 5 min.

was 0.81 g/l. This dose brought about 32% decrease in heart rate, 80% prolongation of the PR interval, 28% decrease in QRS potential, and 43% increase of the QRS duration. The same magnitude of changes in different parameters obtained after oral and intraperitoneal administration of methyl chloroform by different doses is summarized in Table 3.7. This table shows the percent of LC_{50} or of LD_{50} which brought about a fixed percent change in heart rate, PR interval, and QRS potential following inhalation, oral administration, and intraperitoneal injection. Examination of the data in this table reveals that the dose which brings about a 32% decrease in heart rate, 80% increase in the PR interval, and 28% decrease in QRS potential bears a fixed relationship to the lethal amount of each route of administration. For instance, 100% of the LC_{50} determined by inhalation is necessary to bring about the stated quantitative changes in the three parameters. On the other hand, the percent, of the oral LD_{50} which brought about the same changes in the heart rate, PR interval, and QRS potential are 68.2%, 67.5%, and 71.7%, respectively. These values are statistically not significantly different (69.1 ± 1.3). Again, the percent, of the intraperitoneal LD_{50} which brought about the same changes in these parameters are 57.3%, 54.7%, and 59.0%. Here too, these changes are not statistically different (57.0 ± 1.3). The difference between the percent of the lethal dose or concentration (viz., 100% by inhalation, 69.1 ± 1.3% by oral, and 57.0 ± 1.3% by intraperitoneal administration) which brought about the same magnitude of effects on the elctrocardiogram can be explained on the basis of the differences in the absorption, metabolism, and deposition of methyl chloroform given by different routes of administration. However, the consistency of results for each route of administration clearly indicates that each of these electrocardiographic changes bears a fixed relationship to the lethal dose irrespective of the route of administration.

It can also be concluded from the present study that in acute intoxication with methyl chloroform, whether by inhalation in confined spaces or by accidental ingestion, profound changes in cardiovascular function independent of, or perhaps in addition to, depression of the central nervous system are responsible for death.

D. Summary

The lethal amounts of methyl chloroform were determined in male mice after oral, intraperitoneal, and inhalational administration. The single dose that caused 50% mortality after 24 hr was 14.08 ± 2.08 g/kg after oral administration, 2.34 ± 0.32 g/kg after intraperitoneal injection, and 0.81 ± 0.14 g/l (inhaled for 5 min) after inhalation of methyl chloroform. Electrocardiographic studies revealed a dose-dependent decrease in heart rate and in QRS potential and an increase in PR interval and QRS duration. The profound changes in different myocardial parameters independent of, or perhaps in addition to, central nervous system depression are responsible for death. A positive regression correlation between mortality rate and the changes in the electrocardiographic parameters was obtained. The most indicative parameters for toxicity are the decrease in the height of the QRS potential and the increase in its duration.

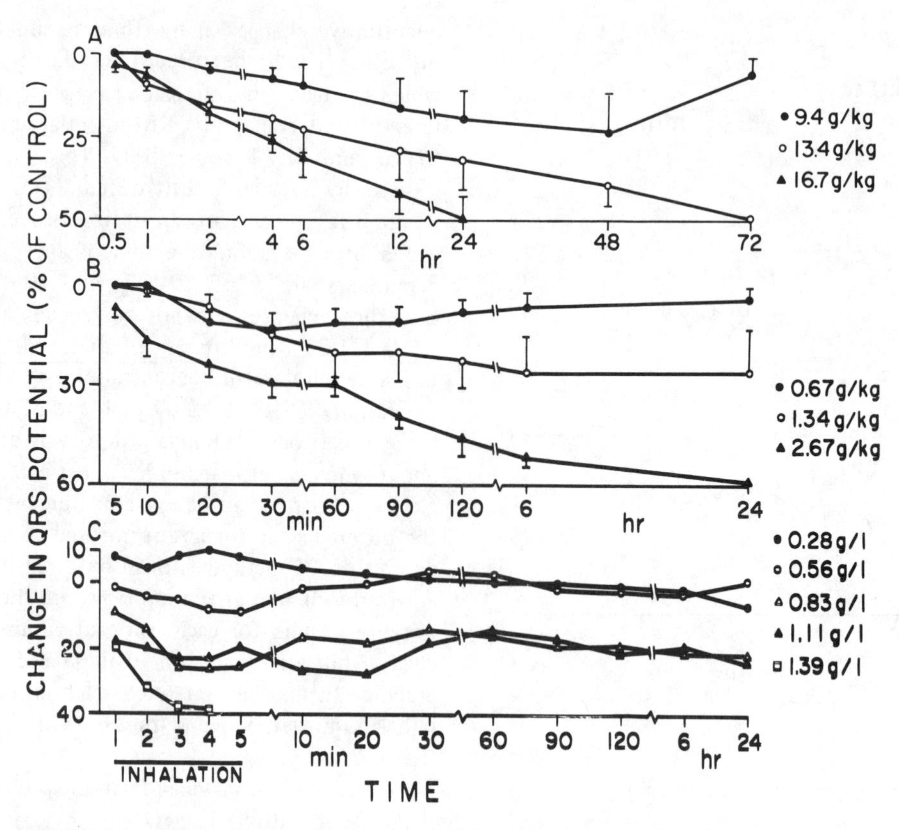

FIGURE 3.4 Effect of various doses of methyl chloroform given orally (A), intraperitoneally (B), or by inhalation (C), on the QRS potential. Note the irreversible decrease in QRS potential elicited by higher doses which led ultimately to death.

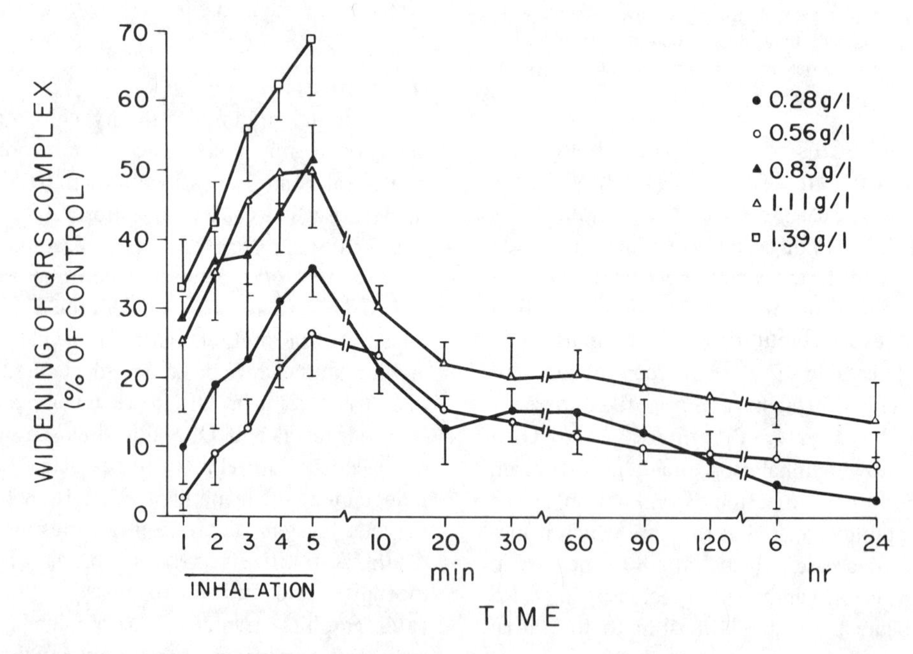

FIGURE 3.5 Effect of various concentrations of methyl chloroform inhaled for 5 min on the duration of QRS complex. Note that high doses brought about a decrease in the intraventricular conduction.

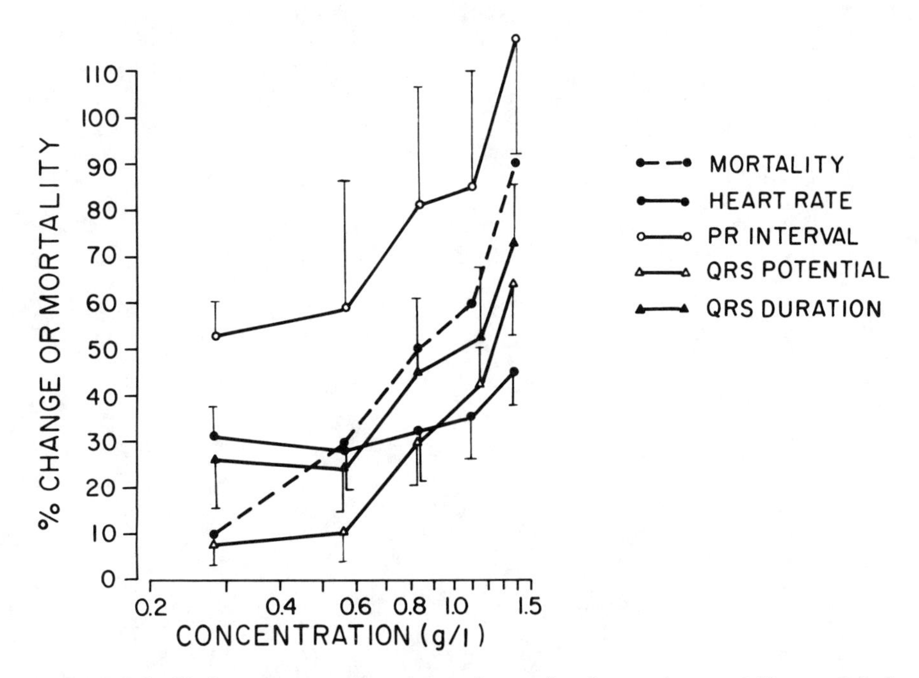

FIGURE 3.6 Maximum response (as percent of control) and percentage mortality are plotted against different concentrations used for inhalation.

TABLE 3.7

Doses that Elicited a Fixed Change in Heart Rate, PR Interval, and QRS Potential Following Inhalation, Oral Administration, and Intraperitoneal Injection of Methyl Chloroform

	Inhalation		Oral		Intraperitoneal	
Parameter (percent change)	Dose g/l	% of LC_{50}	Dose g/kg	% of LD_{50}	Dose g/kg	% of LD_{50}
Decrease in heart rate (32%)	0.81	100	9.6	68.2	1.34	57.3
Increase in PR interval (80%)	0.81	100	9.5	67.5	1.28	54.7
Decrease in QRS potential (28%)	0.81	100	10.1	71.7	1.38	59.0
Mean		100		69.1		57.0
±SEM		±0		±1.3		±1.3

Chapter 4

ACUTE INHALATIONAL TOXICITY OF METHYL CHLOROFORM IN DOGS

The review of the cardiovascular effects of methyl chloroform in Chapter 2 shows that all the available information relates to the cardiotoxicity of this solvent. Specifically, methyl chloroform has been found to alter the electrocardiographic pattern in man and in various animal species,[7,45,75,77] to sensitize the heart to the arrhythmogenic activity of epinephrine[76-80] and to depress myocardial contractile force.[10,11] The present study was undertaken to investigate the effect of varying inhaled concentrations of methyl chloroform on cardiac as well as on systematic vascular and pulmonary vascular functions in anesthetized intact dog preparations. A special effort was made to determine the minimal concentration of methyl chloroform which influences cardiac function, systemic vascular resistance, and pulmonary vascular resistance. Since methyl chloroform is likely to be used with a fluorocarbon propellant in aerosol products, the experiments were designed to determine if there was any interaction between methyl chloroform and one of the more commonly used fluorocarbon inhalants, trichlorofluoromethane (FC 11).

A. Methods

Experiments were performed on seven mongrel dogs of either sex, weighing between 14 and 23 kg (average 18 kg), anesthetized with pentobarbital sodium (30 to 35 mg/kg). After instituting artificial respiration through a tracheostomy with a Starling Ideal respirator, a left thoracotomy was performed in the fourth intercostal space. The pulmonary artery was dissected free from the arch of the aorta and a Statham electromagnetic flow probe was fitted around the artery shortly before its bifurcation. The probe was connected to a Statham SP 2202 electromagnetic flowmeter. Catheters were then placed in the left atrium through the auricular appendage and in the pulmonary artery to the left lower lobe, to measure left artrial and pulmonary arterial pressures, respectively. The cardiac apex was then exposed through a right-sided thoracotomy in the sixth intercostal space and a catheter passed to the left ventricular lumen to measure left ventricular pressure. Aortic pressure was measured by cannulating the right carotid artery. All pressures were measured with P23AA Statham pressure transducers. Recordings were made on a six-channel Sanborn 7700 recorder. After a period of 15 to 20 min during which the preparations were allowed to stabilize, the following concentrations of methyl chloroform, volatilized in air, were administered through the inlet of the respirator for periods of 5 min: 0.05% (2.7 mg/l); 0.1% (5.42 mg/l); 0.25% (13.5 mg/l); 0.5% (27.1 mg/l), and 1% (54.2 mg/l). A period of approximately 10 min was allowed for recovery before a subsequent concentration was administered. In five preparations, 0.5% trichlorofluoromethane (FC 11), previously found to have minimal effects on the cardiovascular system, was administered for a period of 5 min followed, after resumption to inhalation of room air (10 min), by a mixture of 0.5% FC 11 and 0.05% methyl chloroform. The sequence of administration of various concentrations of methyl chloroform, FC 11, and the mixture of FC 11 and methyl chloroform was alternated. Gaseous concentrations were prepared by slowly vaporizing, into a known volume of air, a volume of liquid methyl chloroform or of trichlorofluoromethane (FC 11), calculated to give a gas volume at standard pressure and 20°C appropriate for a particular concentration. Gas chromatography of the mixture revealed that the concentration was within 5% of the estimated level. Data were analyzed by paired comparisons, the criterion for significance being P less than 0.05.

B. Calculations and Abbreviations of Hemodynamic Measurements.

MPAP: Mean pulmonary arterial pressure in cm H_2O, measured from the pulmonary artery to the left lower lobe.

EMPAP: Effective mean pulmonary arterial pressure in cm H_2O, mean pulmonary arterial pressure minus mean left atrial pressure.

MLAP: Mean left atrial pressure in cm H_2O, measured from the left atrium.

LVP: Left ventricular pressure in mm Hg, measured from a catheter in the left ventricular cavity inserted through the cardiac apex.

LVEDP: Left ventricular end-diastolic pressure in mm Hg, measured from the left ventricular pressure.

dp/dt: Maximal rate of rise of left ventricular pressure in mm Hg/sec derived from the left ventricular pressure with a derivative computer.

MAP: Mean aortic pressure in mm Hg, measured from a catheter inserted through a carotid artery.

MPAF: Mean pulmonary arterial flow in ml/min, measured with a Statham electromagnetic flow probe around the main pulmonary artery.

HR: Heart rate, in beats/min, computed from aortic pressure waves, taken at a paper speed of 50 mm/sec.

Systemic vascular resistance: In dynes · sec/cm^5; the quotient of mean aortic pressure minus left ventricular end-diastolic pressure in dynes/cm^2 and of mean pulmonary arterial flow in ml/sec.

Pulmonary vascular resistance: In dynes · sec/cm^5; the quotient of mean pulmonary arterial pressure minus mean left atrial pressure in dynes/cm^2 and of mean pulmonary arterial flow in ml/sec.

C. Results

In general, a dose-related cardiovascular depressant action was exhibited by methyl chloroform. Figure 4.1 shows the record of a typical experiment. Progressively, increasing concentrations of methyl chloroform administered by inhalation led to progressive decreases in peak left ventricular pressure (LVP), maximal rate of rise of left ventricular pressure (dp/dt), aortic pressure (AP), and mean pulmonary arterial flow (MPAF). Table 4.1 summarizes the hemodynamic changes following inhalation of various concentrations of methyl chloroform as well as inhalation of a mixture of subeffective concentrations of trichlorofluoromethane (FC 11) and methyl chloroform. Figure 4.2 shows average per percent changes in some cardiovascular parameters with progressively increasing doses of methyl chloroform.

1. Administration of Methyl Chloroform

Inhalation of a concentration of methyl chloroform of 0.05% for a period of 5 min was without effect on any of the measured parameters. Inhalation of a concentration of 0.1% decreased peak left ventricular pressure, mean aortic pressure, and mean pulmonary arterial flow by 3%, 3%, and 4%, respectively, as compared with control averages (Table 4.1 and Figure 4.2). Inhalation of a concentration of 0.25% decreased peak left ventricular pressure, maximal rate of rise of left ventricular pressure (dp/dt), mean aortic pressure, and mean pulmonary arterial flow by 7%, 13%, 6%, and 6%, respectively, as compared with control averages. A concentration of 0.5% decreased peak left ventricular pressure, maximal rate of rise of left ventricular pressure, (dp/dt), mean aortic pressure, and mean pulmonary arterial flow by 13%, 21%, 14%, and 11%, respectively, as compared with control averages. Following a concentration of 1% there was a decrease in mean left atrial pressure, peak left ventricular pressure, maximal rate of rise of left ventricular pressure (dp/dt), mean aortic pressure, and mean pulmonary arterial flow of 8%, 22%, 33%, 24%, and 15% respectively, as compared with control averages. In addition, systemic vascular resistance decreased by 11% and pulmonary vascular resistance increased by 17%, as compared with control averages. No significant change was noted in the remaining parameters.

2. Administration of a Mixture of Subeffective Concentrations of Methyl Chloroform and Trichlorofluoromethane (FC 11)

The inhalation of 0.5% trichlorofluoromethane, previously found to produce minimal effects in an identical preparation[113] led to a decrease in mean aortic pressure and mean pulmonary arterial flow of 4% and 6%, respectively, as compared with control averages. Administration of a mixture of 0.5% trichlorofluoromethane and 0.05% methyl chloroform led to a decrease in peak left ventricular pressure, mean aortic pressure, and mean pulmonary arterial flow of 4%, 4%, and 6%, respectively, as compared with control averages. There was no significant change in the remaining parameters.

D. Discussion

The results of these experiments indicate that methyl chloroform produced no significant hemodynamic changes in the intact anesthetized open-chest dog preparation in a concentration of 0.05%. However, a concentration of 0.1% decreased peak left ventricular and mean aortic pressures as well as cardiac output. At a concentration of 0.25%, the decreases noted earlier were exaggerated and in addition myocardial contractility, as gauged by the maximal rate of rise of left ventricular pressure (dp/dt), was depressed. Further depression in all these parameters was noted following inhalation of a concentration of 0.5%. Inhalation of a concen-

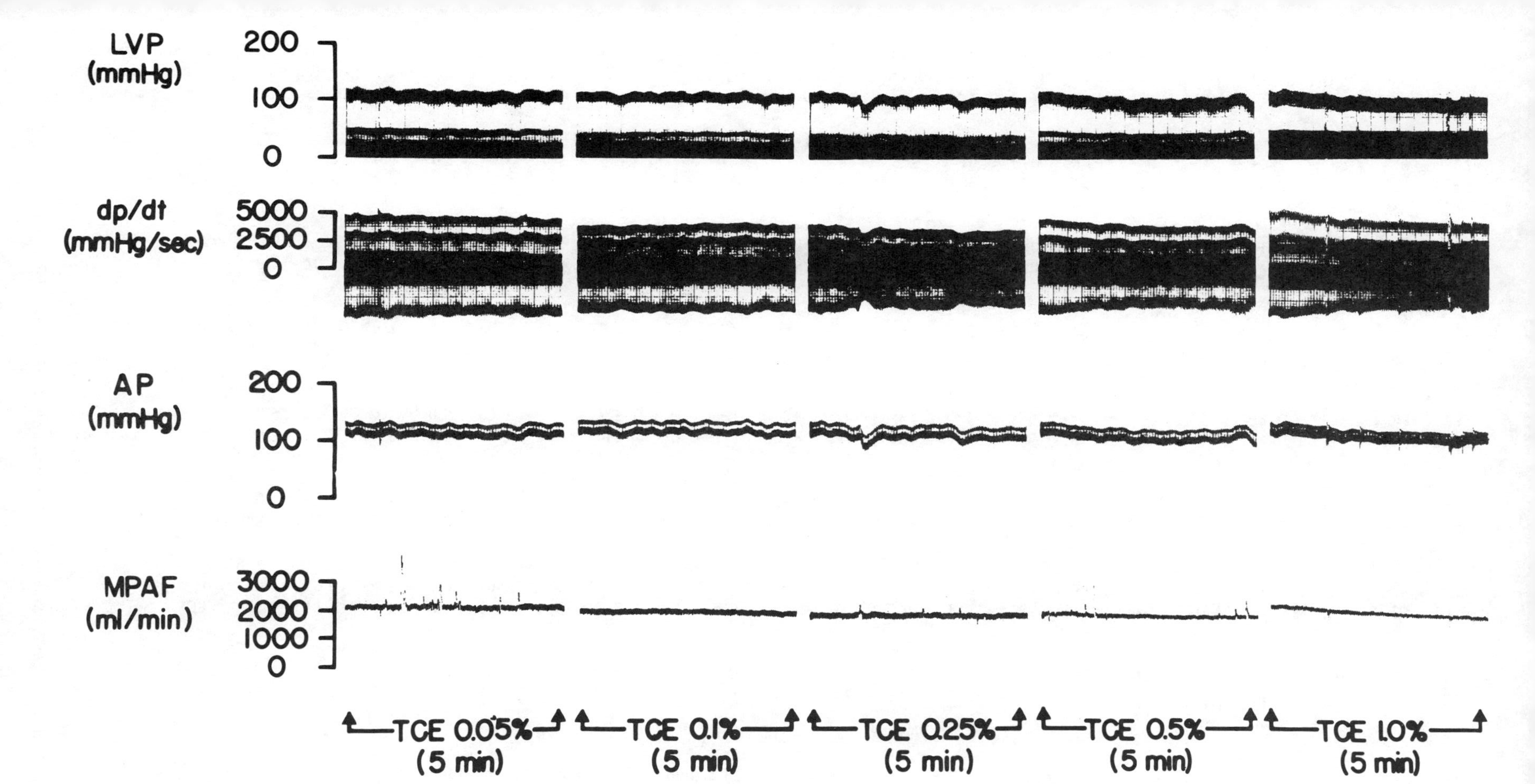

FIGURE 4.1 The effects of progressively increasing concentrations of methyl chloroform on some hemodynamic parameters in a representative experiment. LVP: left ventricular pressure; dp/dt: maximal rate of rise of left ventricular pressure; AP: aortic pressure; MPAF: mean pulmonary arterial flow.

TABLE 4.1

The Cardiovascular Effects of Varying Concentrations of Methyl Chloroform (MC), 0.5% Trichlorofluoromethane (FC 11), and a Mixture of 0.5% FC 11 and 0.05% MC*

	MPAP cmH_2O		EMPAP cmH_2O		MLAP cmH_2O		LVP mmHg		LVEDP mmHg		dp/dt mmHg/sec		MAP mmHg		MPAF ml/min		HR beat/min		Vascular Resistance dynes·sec/cm^5 systemic		Vascular Resistance dynes·sec/cm^5 pulmonary	
	C	E	C	E	C	E	C	E	C	E	C	E	C	E	C	E	C	E	C	E	C	E
MC 0.05%	38.2	38.9	18.4	18.7	19.9	20.2	137	137	3.6	3.6	3136	3064	124	123	1225	1208	151	153	7992	7923	985	1011
	+0.7	±0.4	+0.3	±0.3	+0.3	±0.5	0	±0	0	±0	−72	±71	−1	±0.7	−17	±17	+2	±2	−69	±208	+26	±28
	NS		NS		NS		NS		NS		NS		NS		NS		NS		NS		NS	
MC 0.1%	38.8	38.6	18.9	18.4	19.8	20.2	136	132	4.3	4.3	3175	2975	120	117	1285	1230	150	150	7473	7585	1024	1032
	−0.2	±0.7	−0.5	±0.5	+0.4	±0.4	−4	±1.3	0	±0	−200	±76	−3	±1.0	−55	±16	0	±0	+112	±90	+8	±31
	NS		NS		NS		0.01		NS		NS		0.02		0.01		NS		NS		NS	
MC 0.25%	38.2	38.6	18.0	18.2	20.2	20.3	138	128	5.7	5.7	2986	2593	123	116	1230	1151	163	164	7786	7814	988	1060
	+0.4	±0.7	+0.2	±0.4	+0.1	±0.7	−10	±1.1	0	±0	−393	±75	−7	±1.0	−79	±28	+1	±2	+28	±154	+72	±33
	NS		NS		NS		0.001		NS		0.001		0.001		0.02		NS		NS		NS	
MC 0.5%	38.1	36.8	18.2	17.5	19.8	19.3	135	117	5.7	5.7	2979	2350	122	105	1232	1103	157	157	7480	7264	913	996
	−1.3	±0.9	−0.7	±0.4	−0.5	±0.9	−18	±1.5	0	±0	−629	±80	−17	±1.9	−129	±25	0	±0	−216	±128	+83	±39
	NS		NS		NS		0.001		NS		0.001		0.001		0.001		NS		NS		NS	
MC 1%	36.3	35.4	15.7	16.4	20.6	19.0	131	102	4.3	5.0	2864	1921	119	90	1161	990	142	136	7825	6950	962	1127
	−0.9	±0.6	+0.7	±0.7	−1.6	±0.5	−29	±3.0	+0.7	±0.7	−943	±85	−29	±3.0	−171	±32	−6	±6	−875	±121	+165	±41
	NS		NS		0.02		0.001		NS		0.001		0.001		0.001		NS		0.001		0.01	
FC 11 0.5%	39.8	40.3	20.2	20.6	19.5	19.5	134	129	2.0	2.0	3300	3140	122	117	1230	1160	155	156	7751	7904	991	1091
	+0.5	±0.6	+0.4	±0.5	0	±0	−5	±2.2	0	±0	−160	±93	−5	±1.6	−70	±20	±1	±4	+153	±108	+100	±36
	NS		NS		NS		NS		NS		NS		0.02		0.01		NS		NS		NS	
FC 11 0.5%	38.5	38.5	20.5	20.2	18.0	18.3	125	120	3.0	3.0	2670	2540	114	109	1060	1000	148	148	8523	8664	1179	1288
	+0	±0	−0.3	±0.4	+0.3	±0.5	−5	±0	0	±0	−130	±54	−5	±0	−60	±10	0	±0	+141	±82	+49	±22
MC 0.05%	NS		NS		NS		0.001		NS		NS		0.001		0.001		NS		NS		NS	

Figures represent means of control (C) and experimental (E) data, mean differences ± SE of difference between control and experimental data, and t values. MPAP: mean pulmonary arterial pressure; EMPAP: effective mean pulmonary arterial pressure; MLAP: mean left atrial pressure; LVP: left ventricular pressure; LVEDP: left ventricular end-diastolic pressure; dp/dt: maximal rate of rise of left ventricular pressure; MAP: mean aortic pressure; MPAF: mean pulmonary arterial flow.

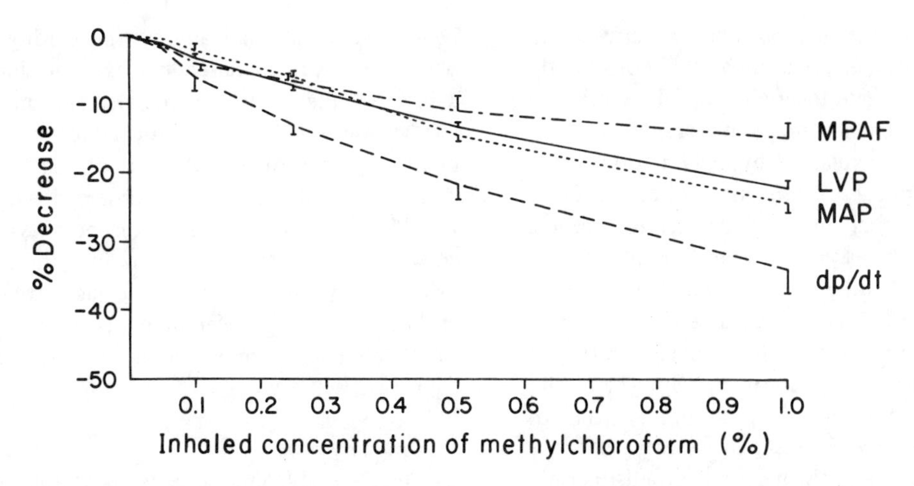

FIGURE 4.2 Mean percentage decrease in left ventricular pressure (LVP), maximal rate of rise of left ventricular pressure (dp/dt), mean aortic pressure (MAP), and mean pulmonary arterial flow (MPAF), in response to inhalation of progressively increasing concentrations of methyl chloroform. Bars indicate SE of mean. Note that at any dose level the decrease in dp/dt is maximal, that in LVP and MAP intermediate, and that in MPAF least.

tration of 1% brought about a marked decrease in the same parameters and in addition it decreased mean left atrial pressure and systemic vascular resistance, while it increased pulmonary vascular resistance.

It may be concluded from the data, therefore, that methyl chloroform exerts a depressant action on the cardiovascular system and that in this regard it is effective in a minimal concentration of 0.1%. At this dose level the most sensitive indices of its cardiovascular depressant effect appear to be its effects on peak left ventricular pressure, mean aortic pressure, and cardiac output, as reflected by mean pulmonary arterial flow. However, as the inhaled concentration of methyl chloroform was increased, depression in the various cardiovascular parameters occurred in varying degrees, being greatest in maximal rate of rise of left ventricular pressure, intermediate in peak left ventricular pressure and mean aortic pressure, and least in cardiac output (Figure 4.2).

The cardiovascular depressant effects of methyl chloroform observed in this study are probably the sum total of effects exerted at more than one site in the cardiovascular system. Identification of the various sites and quantitation of the relative effects of methyl chloroform at each site can only be speculative in view of the fact that the study was performed on an intact preparation. The most outstanding effect of methyl chloroform, in fact an effect which appears with the minimal effective concentration, is a decrease in systemic arterial pressure and in cardiac output. That a direct myocardial depressant action is involved in this effect is evident from the fact that myocardial contractility, as reflected by the maximal rate of rise of left ventricular pressure (dp/dt), decreased most markedly with increasing concentrations of methyl chloroform. Furthermore, depression in myocardial contractility from methyl chloroform has been observed in the perfused frog heart,[75] in the primate heart *in situ,*[10] and in the canine heart-lung preparation.[11] Another possibility which cannot be excluded is a direct relaxant effect on the smooth muscles of the systemic vascular bed, leading to a decrease in vascular resistance and possibly to pooling of blood in the capacitance vessels with consequent decrease in venous return. This possibility is suggested by the fact that, at a concentration of 1%, systemic vascular resistance decreased by 11% as compared with the control average.

The effect of methyl chloroform on the pulmonary component of the cardiovascular system deserves a comment. With inhalation of concentrations of up to 0.5% of methyl chloroform no significant change was observed in mean pulmonary arterial pressure, effective mean pulmonary arterial pressure, mean left atrial pressure, or pulmonary vascular resistance. This occurred at a time when, at all these dose levels, mean pulmonary arterial flow decreased significantly. This apparently paradoxical situation, where the quotient of a constant effective mean pulmonary arterial pressure and a consistently decreasing mean pulmonary arterial flow reflected no significant change in pulmonary vascular resistance, can be explained by examination of the data of individual experiments. Effective mean pulmonary arterial pressure decreased in some experiments, increased in others, and remained constant in still others. Mean pulmonary arterial flow, on the other hand, decreased consistently in all experiments. However, up to a concentration of 0.5% the decrease was maximally 11% and consequently insufficient in order of magnitude to overcome the variations in mean effective pulmonary arterial pressure. In contradistinction, at a dose level of 1%, pulmonary vascular resistance increased significantly by 17%, as compared with the control average. Here too, the variations in effective mean pulmonary arterial pressure occurred to the same extent, so that the average change was insignificant. However, at this dose level mean pulmonary arterial flow decreased drastically by 24%, thus nullifying the less drastic changes in effective mean pulmonary arterial pressure. As a consequence, calculated pulmonary vascular resistance increased. The variations in effective mean pulmonary arterial pressure do not appear to reflect an insignificant effect of methyl chloroform on the pulmonary components of the cardiovascular system as much as the net effect of many forces interacting at the various components of the pulmonary circuit. If methyl chloroform depresses left ventricular contractile function, it is conceivable that it depresses that of the right ventricle equally well. Through this effect as well as a decrease in cardiac output, pulmonary arterial pressure is expected to decrease. A possible direct effect on the pulmonary vascular tree will also be contributory. On the other hand, sympathoadrenal discharge due to the hypotension-induced activation of the carotid-aortic reflex mechanisms produces an opposite effect. The net effect is therefore an algebraic sum of these contradictory forces. Further study is necessary to clarify the effect of methyl chloroform on the various components of the pulmonary circulation.

Administration of a minimal effective concentration of trichlorofluoromethane (FC 11) of 0.5% for a period of 5 min led to a modest but significant decrease in mean aortic pressure and mean pulmonary arterial flow of 4% and 6%, respectively. In a recent publication,[113] this concentration was on the average found to be ineffective in altering any of the cardiovascular parameters, yet in individual experiments a decrease in the same parameters in comparable degrees was noted. However, the addition of 0.05% methyl chloroform to 0.5% FC 11 led to effects indistinguishable from those observed with 0.5% FC 11 alone. This observation permits one to conclude that at such concentrations these agents exhibit no mutual potentiative or additive effects, probably implying different basic mechanisms by which each agent brings about its cardiovascular depressant activity.

E. Summary

The hemodynamic effects of various inhaled concentrations of methyl chloroform were investigated in the anesthetized intact dog preparation. A minimal effective concentration of 0.1% (5.42 mg/l) decreased peak left ventricular pressure, mean aortic pressure, and mean pulmonary arterial flow by 3%, 3%, and 4%, respectively. A concentration of 0.25% (13.9 mg/l) decreased peak left ventricular pressure, maximal rate of rise of left ventricular pressure, dp/dt, mean aortic pressure, and mean pulmonary arterial flow by 7%, 13%, 6%, and 6%, respectively. A concentration of 0.5% (27.1 mg/l) exaggerated all previous responses. A concentration of 1% (54.2 mg/l) decreased peak left ventricular pressure, dp/dt, mean aortic pressure, mean pulmonary arterial flow, and systemic vascular resistance by 22%, 33%, 24%, 15%, and 11%, respectively, while it increased pulmonary vascular resistance by 17%. It may be concluded from this study that methyl chloroform is a general depressant of cardiovascular function, effective in a minimal inhaled concentration of 0.1%. At this concentration the most sensitive indices of its cardiovascular depressant effects are peak systolic pressure, mean aortic pressure, and cardiac output. With progressively increasing concentrations the decrease in the maximal rate of rise of left ventricular pressure, dp/dt, occurs to the greatest extent, while the decrease in cardiac output occurs to the least extent, the decrease in mean aortic pressure and in peak left ventricular pressure occupying an intermediate position. A mixture of 0.5% FC 11 and 0.05% methyl chloroform exhibits no potentiative or additive effects, probably implying differences in basic mechanisms by which each agent brings about its cardiovascular depressant action. The effects demonstrated in these experiments serve to explain death from poisoning with methyl chloroform.

Chapter 5

SUMMARY OF TOXICITY OF METHYL CHLOROFORM

This monograph on methyl chloroform was prepared to identify the health hazards, if any, of methyl chloroform when used in aerosol products. It has become apparent that additional investigation is needed to determine if any health hazard exists when aerosol products are properly used.

There is sufficient information available to specify what organs are affected when individuals suffer from methyl chloroform poisoning. The experiments in man and animals are summarized in Table 5.1. The organs which are influenced by methyl chloroform include the heart, lungs, liver, kidneys, and central nervous system.

The heart is the organ most vulnerable to the exposure to methyl chloroform, since a significant depression of contractility can be elicited in the dog with exposure to 5.6 mg/l. The minimal concentration that would depress respiration has not been determined in the dog or in the monkey, although in the guinea pig chronic exposure (1 hr daily for 3 months) to 5.5 mg/l causes lesions in the lungs. The susceptibility of the liver varies between animals and man; inhalation of a concentration of 14.5 mg/l (for 15 min) influences the liver in man but in mice the hepatotoxic dose is close to the lethal inhalation dose. There are no threshold studies relating to the kidneys.

The susceptibility of the central nervous system has not been investigated in the dog. Although adequate amounts would produce anesthesia, there is no investigation on the threshold level that would produce disequilibrium so that it is not possible to compare the effects on the heart and the central nervous system. In man, the level that produces disequilibrium is 10 mg/l (5 min)[36] and 15 mg/l (15 min),[41] which is two to three times the threshold level for cardiotoxicity in the dog.

The experiments described in Chapters 2 and 3 are helpful in identifying the pattern of cardiotoxicity when sufficient concentrations of methyl chloroform are inhaled. The effects on the electrocardiogram, myocardial contractility, cardiac output, pulmonary vascular resistance, and systemic vascular resistance in experimental animals have been elucidated. The threshold concentrations for the various parameters have been established when methyl chloroform is administered acutely for brief periods. The next step is to administer this solvent repeatedly every day for an extended period of time to determine if the acute responses continue to appear in an exaggerated fashion or disappear as a result of the development of tolerance. Investigation along these lines will ultimately lead to the understanding of the effects of the presence of methyl chloroform in the environment resulting from its use in consumer products as well as in industry.

TABLE 5.1

Summary of Available Information on the Toxicity of Methyl Chloroform

	Mice	Rats	Guinea Pigs	Dogs	Monkeys	Humans
Lethal Dose	Oral LD_{50} = 9.7 – 11.2 g/kg[36]	Oral LD_{50} = 10 –14.3 g/kg[36]	Oral LD_{50} = 8.6 – 9.5 g/kg[36]	i.p. LD_{50} = 4.14 g/kg[65]		Inhal LD = 27.5 mg/l[36] blood level in fatalities 13 and 72 mg/100 ml[30]
	Oral LD_{50} = 14.08 g/kg* i.p. LD_{50} = 4.7 –4.9 g/kg[62,63] i.p. LD_{50} = 2.34 g/kg*	i.p. LD_{50} = 5.08 g/kg[64]				
	Inhal LD_{50} = 65 mg/l[74]	Inhal LD_{50} = 98 mg/l (3 hr) and 76 mg/l (7 hr)[60]				
	Inhal LD_{50} = 810 mg/l (5 min)*					
Cardio-toxicity	Inhal 900 mg/l = no arrhythmia and no sensitization to epinephrine.[7]			i.v. 0.33 –0.53 g/kg = arrhythmia[76]	Inhal 138.8 mg/l = depress contractility and hypotension.[10]	Inhal 33.3 – 124.9 mg/l = cardiac arrhythmia[45]
	Inhal 280 mg/l (5 min) = cardiac depression*			Inhal 27.8 mg/. = sensitize to epinephrine[77]		
				Inhal 277 mg/l = depress contractility[11] Inhal 2.8 mg/l = no effect* Inhal 15.6 mg/l = depress		

*Based on experiments reported in Chapters 3 and 4.

TABLE 5.1 (continued)

Summary of Available Information on the Toxicity of Methyl Chloroform

	Mice	Rats	Guinea Pigs	Dogs	Monkeys	Humans
Pneumo-toxicity			Inhal 15.5 mg/l (1 hr daily for 3 mo) = lung lesions[36]	Inhal 800 mg/kg = respiratory failure[75]	Inhal 138.8 mg/l = respiratory depression and bronchodilation[14]	
Hepato-tocicity	i.p. hepatotoxic ED_{50} = 11.2 g/kg[85]	Oral 1.65 g/kg (daily 7 days) = increased hepatic enzymes[94]	Inhal 27.3 mg/l (7 hr daily for 10 weeks) = fatty liver[60]		Inhal 16.4 mg/l (7 hr daily for 10 weeks) = no effect[60]	Inhal 2.75 mg/l = no effect[41]
	i.p. hepatotoxic ED_{50} = 3.34 g/kg[63]	i.p. 3.74 g/kg = no effect[64]	Inhal 5.5 mg/l (1.2 hr daily for 3 mo) = liver pathology[36]		Inhal 2.75 mg/l (7 hr daily for 6 mo) = no effect[36]	Inhal 14.5 mg/l (15 min) = decreased urobilinogen[41]
	Inhal ED_{50} = 74.3 mg/l (595 min)[62]	Inhal 166 mg/l (7 hr) increased liver weight[60]	inhal 12.1 mg/l (8 hr daily for 30 days) = no effect[67]			
	Inhal 16.5 mg/l = increased hepatic enzyme[87]	Inhal 44 mg/l (7 hr) = fatty liver[60]				
		Inhal 55 mg/l (1hr daily for 3 mo) = increased liver weight[36]				
		Inhal 13.75 – 16.5 mg/l (24 hr) = increased hepatic enzymes[88]				

TABLE 5.1 (continued)

Summary of Available Information on the Toxicity of Methyl Chloroform

	Mice	Rats	Guinea Pigs	Dogs	Monkeys	Humans
Nephro-toxicity		Inhal 98 mg/l (12 hr) = increased kidney weight[60]				
Central Nervous System Depression	Inhal 45 mg/l = narcosis[74]	Inhal 98 mg/l (0.3 hr); 55 mg/l (3 hr) or 44 mg/l (7 hr) = depression[60]		Inhal 0.45 g/kg = anesthesia[75]		Inhal 33.3 – 124.9 mg/l = anesthesia[45] Inhal 2.48 mg/l (4 hr) = no behavioral effect[43] Inhal 2.75 mg/l (7 hr) = positive Romberg[42] Inhal 2.75 mg/l (7 hr daily for 5 days) = no effect[69] Inhal <5.5 mg/l (15 min) = smell[41] Inhal 5.5 – 6.05 mg/l (15 min) = eye irritation[41] Inhal 10.45 – 11.0 mg/l (15 min) = throat irritation[41] Inhal 14.3 – 15 mg/l (15 min) = lighthead[41] Inhal 14.8 mg/l (15 min) = disequilibrium[41] Inhal 2.48 – 3.9 mg/l (90 min) = no smell[36] Inhal 4.9 – 6.5 mg/l (30 min) = unique odor[36] Inhal 4.9 – 5.5 mg/l (75 min) = eye irritation and lightheadedness[36] Inhal 9.57 – 11.99 mg/l (5 min) = disequilibrium[36]

REFERENCES – Methyl Chloroform

1. **Aviado, D. M.,** *Sympathomimetic drugs,* Charles C Thomas, Springfield, Illinois, 1970, chap. 11.
2. **Aviado, D. M.,** Cardiopulmonary effects of fluorocarbon compounds, in *Proceedings of the 2nd Annual Conference on Environmental Toxicology,* Aerospace Medical Research Laboratory, Wright-Patterson Air Force Base, Dayton, Ohio, 1971, 31.
3. **Aviado, D. M.,** Kratschmer reflex induced by hazards of halocarbon propellants, in *Conference on Toxic Hazards of Halocarbon Propellants,* Thompson, G.E., Ed., Dept. of Health, Education, and Welfare, Public Health Service, Food and Drug Administration, Washington, D.C., 1972, 63.
4. **Aviado, D. M.,** Toxicity of aerosols, *J. Clin. Pharmacol.,* 15, 86, 1975.
5. **Aviado, D. M.,** Toxicity of propellants, in *Proceedings of the 4th Annual Conference on Environmental Toxicology,* Aerospace Medical Research Laboratory, Wright-Patterson Air Force Base, Dayton, Ohio, 1973, 291.
6. **Aviado, D. M.,** Toxicity of propellants, in *Progress in Drug Research,* Jucker, E., Ed., Birkhäuser Verlag, Basel, 1974, 365.
7. **Aviado, D. M. and Belej, M. A.,** Toxicity of aerosol propellants on the respiratory and circulatory systems. I. Cardiac arrhythmia in the mouse, *Toxicology,* 2, 31, 1974.
8. **Friedman, S. A., Cammarato, M., and Aviado, D. M.,** Toxicity of aerosol propellants on the respiratory and circulatory systems. II. Respiratory and bronchopulmonary effects in the rat, *Toxicology,* 1, 345, 1973.
9. **Brody, R. S., Watanabe, T., and Aviado, D. M.,** Toxicity of aerosol propellants on the respiratory and circulatory systems. III. Influence of bronchopulmonary lesion on cardio-pulmonary toxicity in the mouse, *Toxicology,* 2, 173, 1974.
10. **Belej, M. A., Smith, D. G., and Aviado, D. M.,** Toxicity of aerosol propellants in the respiratory and circulatory systems. IV. Cardiotoxicity in the monkey, *Toxicology,* 2, 381, 1974.
11. **Aviado, D. M. and Belej, M. A.,** Toxicity of aerosol propellants in the respiratory and circulatory systems. V. Ventricular function in the dog, *Toxicology,* 3, 79, 1975.
12. **Doherty, R. E. and Aviado, D. M.,** Toxicity of aerosol propellants in the respiratory and circulatory systems. VI. Influence of cardiac and pulmonary vascular lesions in the rat, *Toxicology,* 3, 213, 1975.
13. **Watanabe, T. and Aviado, D. M.,** Toxicity of aerosol propellants in the respiratory and circulatory systems. VII. Influence of pulmonary emphysema and anesthesia in the rat, *Toxicology,* 3, 225, 1975.
14. **Aviado, D. M. and Smith, D. G.,** Toxicity of aerosol propellants in the respiratory and circulatory systems. VIII. Respiration and circulation in primates, *Toxicology,* 3, 241, 1975.
15. **Aviado, D. M.,** Toxicity of aerosol propellants in the respiratory and circulatory systems. IX. Summary of the most toxic: trichlorofluoromethane (FC11), *Toxicology,* 3, 311, 1975.
16. **Aviado, D. M.,** Toxicity of aerosol propellants in the respiratory and circulatory systems. X. Proposed classification, *Toxicology,* 3, 321, 1975.
17. **Belej, M. A. and Aviado, D. M.,** Cardiopulmonary toxicity of propellants for aerosols, *J. Clin. Pharmacol.,* 15, 105, 1975.
18. **Aviado, D. M. and Drimal, J.,** Five fluorocarbons for administration of aerosol bronchodilators, *J. Clin. Pharmacol.,* 15, 116, 1975.
19. **Schober, A. E.,** Chlorothene in hair sprays, *Soap Chem. Spec.,* 34, 65, 1958.
20. **Stewart, R. D.,** The toxicology of methyl chloroform, *J. Occup. Med.,* 5, 259, 1963.
21. **Stewart, R. D. and Rowe, V. K.,** Quinze ans d'etudes sur le 1,1,1-trichloroéthane, *Arch. Mal. Prof. Med. Trav. Secur. Soc.,* 28, 194, 1967.
22. **Stewart, R. D.,** The toxicology of 1,1,1-trichloroethane, *Ann. Occup. Hyg.,* 11, 71, 1968.
23. **Stewart, R. D.,** The toxicology of 1,1,1-trichloroethane, *Med. Lav.,* 59, 6, 1968.
24. **Blankenship, M. J.,** The aerothene propellant systems, *Aerosol Rep.,* 7, 132, 1968.
25. **Chenoweth, M. B. and Hake, C. L.,** The smaller halogenated aliphatic hydrocarbons, *Annu. Rev. Pharmacol.,* 2, 363, 1962.
26. **Crummett, W. B. and Stenger, V. A.,** Thermal stability of methyl chloroform and carbon tetrachloride, *Ind. Eng. Chem.,* 48, 434, 1956.
27. **Sjoberg, B.,** *Sven. Kem. Tidskr.,* 64, 1972. Cited from Crummett and Stenger in Reference 26 above.
28. **Anthony, T.,** New propellant systems based on chlorinated solvents and compressed gases, *Aerosol Age,* 129, 31, 1967.
29. **Parker, K. D., Yee, J. L., and Kirk, P. L.,** Gas chromatographic determination of ethyl alcohol in blood for medicolegal purposes: separation of other volatiles from blood or aqueous solution, *Anal. Chem.,* 34, 1234, 1962.
30. **Hall, F. B. and Hine, C. H.,** Trichloroethane intoxication: a report of two cases, *J. Forensic Sci.,* 11, 404, 1966.
31. **Barber, J. V.,** Chlorothene in aerosols, *Soap Chem. Spec.,* 99, 209, 1957.
32. Trichloroethane(1,1,1-trichloroethane, methyl chloroform), *Fed. Regist.,* 38, 21935, August 14, 1973.
33. Human drugs containing tricholorethane (1,1,1-tricholoroethene, methylchloroform), *Fed. Regist.,* 38, 21949, August 15, 1973.
34. **Gleason, M. N., Gosselin, R. E., Hodge, H. C., and Smith, R. P.,** *Clinical Toxicology of Commercial Products,* 3rd ed., Williams and Wilkins, Baltimore, 1969.

35. **Kleinfeld, M. and Feiner, B.**, Health hazards associated with work in confined spaces, *J. Occup. Med.*, 8, 358, 1966.
36. **Torkelson, T. R., Oyen, F., McCollister, D. D., and Rowe, V. K.**, Toxicity of 1,1,1-trichloroethane as determined on laboratory animals and human subjects, *Am. Ind. Hyg. Assoc. J.*, 19, 353, 1958.
37. **Stahl, C. J., Fatteh, A. V., and Dominguez, A. M.**, Tricholoroethane poisoning: observations on the pathology and toxicology in six fatal cases, *J. Forensic Sci.*, 14, 393, 1969.
38. **Hatfield, T. R. and Maykoski, R. T.**, A fatal methyl chloroform (trichloroethane) poisoning, *Arch. Environ. Health*, 20, 279, 1970.
39. **Stewart, R. D. and Andrews, J. T.**, Acute intoxication with methyl chloroform, *J. Am. Med. Assoc.*, 195, 904, 1966.
40. **Travers, H.**, Death from 1,1,1-trichloroethane abuse: case report, *Mil. Med.*, 139, 889, 1974.
41. **Stewart, R. D., Gay, H. H., Erley, D. S., Hake, C. L., and Schaffer, A. W.**, Human exposure to 1,1,1-trichloroethane vapor: relationship of expired air and blood concentrations to exposure and toxicity, *Am. Ind. Hyg. Assoc. J.*, 22, 252, 1961.
42. **Stewart, R. D., Gay, H. H., Schaffer, A. W., Erley, D. S., and Rowe, V. K.**, Experimental human exposure to methyl chloroform vapor, *Arch. Environ. Health*, 19, 467, 1969.
43. **Salvini, M., Binaschi, S., and Riva, M.**, Evaluation of the psychophysiological functions in humans exposed to the 'threshold limit value' of 1,1,1-trichloroethane, *Br. J. Ind. Med.*, 28, 286, 1971.
44. **Gazzaniga, G., Binaschi, S., Sportelli, A., and Riva, M.**, L'eliminazione nell'aria alveolare dell'uomo dell' 1,1,1-tricloroetano dopo esposizione a 600 ppm per 3 ore., *Boll. Soc. Ital. Biol. Sper.*, 45, 97, 1969.
45. **Dornette, W. H. L. and Jones, J. P.**, Clinical experiences with 1,1,1 trichloroethane; a preliminary report of 50 anesthetic administrations, *Anesth. Analg.* (Cleveland), 39, 249, 1960.
46. **Simmons, J. H. and Moss, I. M.**, Measurement of personal exposure to l,l,l-trichloroethane and trichloroethylene using an inexpensive sampling device and battery-operated pump, *Ann. Occup. Hyg.*, 15, 47, 1973.
47. **Astrand, I., Kilbom, A., Wahlberg, I., and Ovrum, P.**, Methylchloroform exposure. I. Concentration in alveolar air and blood at rest and during exercise, *Work Environ. Health*, 10, 69, 1973.
48. **Gamberale, F. and Hultengren, M.**, Methylchloroform exposure. II. Psychophysiological functions, *Work Environ. Health*, 10, 82, 1973.
49. **Morgan, A., Black, A., and Belcher, D. R.**, The excretion in breath of some aliphatic halogenated hydrocarbons following administration by inhalation, *Ann. Occup. Hyg.*, 13, 219, 1970.
50. **Morgan, A., Black, A., and Belcher, D. R.**, Studies on the absorption of halogenated hydrocarbons and their excretion in breath using ^{38}Cl tracer techniques, *Ann. Occup. Hyg.*, 15, 273, 1972.
51. **Tada, O., Nakaaki, K., and Fukabori, S.**, On the methods of determinations of chlorinated hydrocarbons in the air and their metabolisms in the urine, *Rodo Kagaku*, 44, 500, 1968.
52. **Monzani, C., Rasetti, L., and De Pedrini, C.**, Aspetti tossicologici dell'1-1-1-trichloroethano, *Arch. Sci. Med.*, 125, 777, 1969.
53. **Seki, Y., Urashima, Y., Aikawa, H., Matsumura, H., Ichikawa, Y., Hiratsuka, F., Yoshioka, Y., Shimbo, S., and Ikeda, M.**, Trichloro-compounds in the urine of humans exposed to methyl chloroform at sub-threshold levels, *Int. Arch. Arbeitsmed.*, 34, 39, 1975.
54. **Hake, C. L., Waggoner, T. B., Robertson, D. N., and Rowe, V. K.**, The metabolism of 1,1,1-trichloroethane by the rat, *Arch. Environ. Health*, 1, 101, 1960.
55. **Boettner, E. A. and Muranko, H. J.**, Animal breath data for estimating the exposure of humans to chlorinated hydrocarbons, *Am. Ind. Hyg. Assoc. J.*, 39, 437, 1969.
56. **Ikeda, M. and Ohtsuji, H.**, A comparative study of the excretion of Fujiwara reaction– positive substances in urine of humans and rodents given trichloro- or tetrachloro-derivatives of ethane and ethylene, *Br. J. Ind. Med.*, 29, 99, 1972.
57. **Eben, A. and Kimmerle, G.**, Metabolism, excretion and toxicology of methylchloroform in acute and subacute exposed rats, *Arch. Toxikol.*, 31, 233, 1974.
58. **Patty, F. A.**, *Industrial Hygiene and Toxicology*, Interscience, New York, 1949, 807.
59. **Carpenter, C. P., Smyth, H. F., Jr., and Pozzani, U. C.**, The assay of acute vapor toxicity and the grading and interpretation of results on 96 chemical compounds, *J. Ind. Hyg. Toxicol.*, 31, 343, 1949.
60. **Adams, E. M., Spencer, H. C., Rowe, V. K., and Irish, D. D.**, Vapor toxicity of 1,1,1-trichloroethane (methylchloroform) determined by experiments on laboratory animals, *Arch. Ind. Hyg. Occup. Med.*, 1, 225, 1950.
61. **Takeuchi, Y.**, Experimental studies on the toxicity of 1,1,1,2-tetrachloroethane compared with 1,1,2,2-tetrachloroethane and 1,1,1-trichloroethane, *Jpn. J. Ind. Health*, 8, 371, 1966.
62. **Gehring, P. J.**, Hepatotoxic potency of various chlorinated hydrocarbon vapours relative to their narcotic and lethal potencies in mice, *Toxicol. Appl. Pharmacol.*, 13, 287, 1968.
63. **Klaassen, C. D. and Plaa, G. L.**, Relative effects of various chlorinated hydrocarbons on liver and kidney function in mice, *Toxicol. Appl. Pharmacol.*, 9, 139, 1966.
64. **Klaassen, C. D. and Plaa, G. L.**, Comparison of the biochemical alterations elicited in livers from rats treated with carbon tetrachloride, chloroform, 1,1,2-trichloro-ethane and 1,1,1-trichloroethane, *Biochem. Pharmacol.*, 18, 2019, 1969.
65. **Klaassen, C. D. and Plaa, G. L.**, Relative effects of various chlorinated hydrocarbons on liver and kidney function in dogs, *Toxicol. Appl. Pharmacol.*, 10, 119, 1967.

66. **Siegel, J., Jones, R. A., Coon, R. A., and Lyon, J. P.,** Effects on experimental animals of acute, repeated and continuous inhalation exposures to dichloroacetylene mixtures, *Toxicol. Appl. Pharmacol.,* 18, 168, 1971.
67. **Prendergast, J. A., Jones, R. A., Jenkins, L. J., Jr., and Siegel, J.,** Effects on experimental animals of long-term inhalation of trichloroethylene, carbon tetrachloride, 1,1,1-trichloroethane, dichlorodifluoromethane, and 1,1-dichloroethylene, *Toxicol. Appl. Pharmacol.,* 10, 270, 1967.
68. **Tsapko, V. G. and Rappoport, M. E.,** Effect of methylchloroform vapors on animals, *Farmakol. Toksikol.* (Moscow), 7, 149, 1972.
69. **Rowe, V. K., Wujkowski, T., Wolf, M. A., Sadek, S. E., and Stewart, R. D.,** Toxicity of a solvent mixture of 1,1,1-trichloroethane and tetrachloroethylene as determined by experiments on laboratory animals and human subjects, *Am. Ind. Hyg. Assoc. J.,* 24, 541, 1963.
70. **Llewellyn, O. P.,** Halogenated hydrocarbons used as solvents, *Ann. Occup. Hyg.,* 9, 199, 1966.
71. **Von Oettingen, W. F.,** The halogenated hydrocarbons: their toxicity and potential dangers, *J. Ind. Hyg. Toxicol.,* 19, 349, 1937.
72. **Von Oettingen, W. F.,** Common industrial solvents and their systemic effects, *Conn. State Med. J.,* 8, 485, 1944.
73. **Von Oettingen, W. F.,** *The Halogenated Hydrocarbons of Industrial and Toxicological Importance,* Elsevier, Amsterdam, 1964.
74. **Lazarew, N. W.,** Über die narkotische Wirkungskraft der Dämpfe der Chlorderivaten des Methans, des Äthans und des Äthylens, *Arch. Exp. Path. Pharmakol.,* 141, 19, 1929.
75. **Krantz, J. C., Jr., Park, C. S., and Ling, J. S. L.,** Anesthesia. LX: The anesthetic properties of 1,1,1-trichloroethane, *Anesthesiology,* 20, 635, 1959.
76. **Rennick, B. R., Malton, S. D., Moe, G. K., and Seevers, M. H.,** Induction of idioventricular rhythms by 1,1,1-trichloroethane and epinephrine, *Fed. Proc.,* 8, 327, 1949.
77. **Reinhardt, C. F., Mullin, L. S., and Maxfield, M. E.,** Epinephrine-induced cardiac arrhythmia potential of some common industrial solvents, *J. Occup. Med.,* 15, 953, 1973.
78. **Clark, D. G. and Tinston, D. J.,** Correlation of the cardiac sensitizing potential of halogenated hydrocarbons with their physiochemical properties, *Br. J. Pharmacol.,* 49, 355, 1973.
79. **Somani, P. and Lum, B. K. B.,** The antiarrhythmic actions of beta adrenergic blocking agents, *J. Pharmacol. Exp. Ther.,* 147, 194, 1965.
80. **Lucchesi, B.,** The effects of pronethalol and its dextro isomer upon experimental cardiac arrhythmias, *J. Pharmacol. Exp. Ther.,* 148, 94, 1965.
81. **Hermansen, K.,** Antifibrillatory effect of some β-adrenergic receptor blocking agents determined by a new test procedure in mice, *Acta Pharmacol. Toxicol.,* 28, 17, 1970.
82. **Strosberg, A. M., Johnson, L., and Miller, L.,** Methyl chloroform induced fibrillation in the mouse, *Fed. Proc.,* 32, 3178, 1973.
83. **Herd, P. A., Lipsky, M., and Martin, H. F.,** Cardiovascular effects of 1,1,1-trichloroethane, *Arch. Environ. Health,* 28, 227, 1974.
84. **Truhaut, R., Boudène, C., Jouany, J.-M., and Bouant, A.,** Application du physiogramme à l'étude de la toxicologie aiguë des solvants chlorés, *Eur. J. Toxicol.,* 5, 284, 1972.
85. **Plaa, G. L., Evans, E. A., and Hine, C. H.,** Relative hepatotoxicity of seven halogenated hydrocarbons, *J. Pharmacol. Exp. Ther.,* 123, 224, 1958.
86. **MacEwen, J. D., Kinkead, E. R., and Haun, C. C.,** A study of the biological effect of continuous inhalation exposure of 1,1,1-trichloroethane (methyl chloroform) on animals, *NASA Contract, Rep.,* 134323, 1974.
87. **Lal, H. and Shah, H. C.,** Effect of methylchloroform inhalation on barbiturate hypnosis and hepatic drug metabolism in male mice, *Toxicol. Appl. Pharmacol.,* 17, 625, 1970.
88. **Fuller, G. C., Olshan, A., Puri, S. K., and Lal, H.,** Induction of hepatic drug metabolism in rats by methylchloroform inhalation, *J. Pharmacol. Exp. Ther.,* 175, 311, 1970.
89. **Cornish, H. H. and Adefuin, J.,** Ethanol potentiation of halogenated aliphatic solvent toxicity, *Am. Ind. Hyg. Assoc. J.,* 27, 57, 1966.
90. **Cornish, H. H., Ling, B. P., and Barth, M. L.,** Phenobarbital and organic solvent toxicity, *Am. Ind. Hyg. Assoc. J.,* 34, 487, 1973.
91. **Hanasono, G. K., Witschi, H., and Plaa, G. L.,** Potentiation of the hepatotoxic responses to chemicals in alloxan-diabetic rats, *Proc. Soc. Exp. Biol. Med.,* 149, 903, 1975.
92. **Rice, A. J., Roberts, R. J., and Plaa, G. L.,** The effect of carbon tetrachloride, administered *in vivo*, on the hemodynamics of the isolated perfused rat liver, *Toxicol. Appl. Pharmacol.,* 11, 422, 1967.
93. **Herd, P. A. and Martin, H. F.,** Effect of 1,1,1-trichloroethane on mitochondrial metabolism, *Biochem. Pharmacol.,* 24, 1179, 1975.
94. **Platt, D. S. and Cockrill, B. L.,** Biochemical changes in rat liver in response to treatment with drugs and other agents. II. Effects of halothane, DDT, other chlorinated hydrocarbons, thioacetamide, dimethylnitrosamine and ethionine, *Biochem. Pharmacol.,* 18, 445, 1969.
95. **Truhaut, R., Boudene, C., Phu-lich, N., and Catella, H.,** La determination de certaines activities enzymatiques seriques comme test d'agressivite hepatique de divers solvants chlores industriels chez le lapin, *Arch. Mal. Prof. Med. Trav. Secur. Soc.,* 28, 425, 1967.

96. **Lucas, G. H. W.,** A study of the fate and toxicity of bromine and chlorine containing anesthetics, *J. Pharmacol. Exp. Ther.,* 34, 223, 1928.
97. **Barrett, H. M., Cunningham, J. G., and Johnston, J. H.,** A study of the fate in the organism of some chlorinated hydrocarbons, *J. Ind. Hyg. Toxicol.,* 21, 479, 1939.
98. **Plaa, G. L. and Larson, R. E.,** Relative nephrotoxic properties of chlorinated methane, ethane, and ethylene derivatives in mice, *Toxicol. Appl. Pharmacol.,* 7, 37, 1965.
99. **Marzulli, F. N. and Ruggles, D. I.,** Rabbit eye irritation test: collaborative study, *J. Assoc. Off. Anal. Chem.,* 56, 905, 1973.
100. **Stewart, R. D. and Dodd, H. C.,** Absorption of carbon tetrachloride, trichloroethylene, tetrachloroethylene, methylene chloride, and 1,1,1-trichloroethane through the human skin, *Am. Ind. Hyg. Assoc. J.,* 25, 439, 1964.
101. **Leong, B. K. J., Schwetz, B. A., and Gehring, P. J.,** Embryo and fetotoxicity of inhaled trichloroethylene, perchloroethylene, methyl chloroform and methylene chloride in mice and rats, Abstract of Papers, Society of Toxicology, 14th Annual Meeting, March 1975, 29.
102. **Schwetz, B. A., Leong, B. K. J., and Gehring, P. J.,** The effect of maternally inhaled trichloroethylene, perchloroethylene, methyl chloroform and methylene chloride on embryonal and fetal development in mice and rats, *Toxicol. Appl. Pharmacol.,* 32, 84, 1975.
103. **Binaschi, S., Gazzaniga, G., Rizzo, S., and Riva, M.,** Integrazione fra spettrografia all'infrarosso e gascromatografia in un caso di intossicazione da alogeno-derivati degli idrocarburi, *Boll. Soc. Ital. Biol. Sper.,* 45, 94, 1969.
104. **Stewart, R. D.,** Methyl chloroform intoxication; diagnosis and treatment, *J. Am. Med. Assoc.,* 215, 1789, 1971.
105. **Stokinger, H. E., Ashe, H. B., Baier, E. J., Coleman, A. L., Elkins, H. B., Grabois, B., Hayes, W. J., Jr., Jacobson, K. H., MacFarland, H. N., Reindollar, W. F., Scovill, R. G., Smith, R. G., and Zavon, M. R.,** Threshold limit values for 1963, *J. Occup. Med.,* 5, 491, 1963.
106. American Industrial Hygiene Association, Toxicology Committee, Emergency exposure limits, *Am. Ind. Hyg. Assoc. J.,* 25, 578, 1964.
107. **Luxon, S. G.,** Recent developments in the use of solvents, *Ann. Occup. Hyg.,* 9, 231, 1966.
108. American Conference of Governmental Industrial Hygienists, Documentation of the threshold limit values for substances in workroom air, Cincinnati, Ohio, 1971, 161.
109. **Schaffer, A. W.,** Comparative toxicity of chlorinated solvents, *Arch. Mal. Prof. Med. Trav. Secur. Soc.,* 31, 150, 1970.
110. **Patty, F. A.,** *Industrial Hygiene and Toxicology,* Vol. II, 2nd ed., Interscience, New York, 1963, 1288.
111. **Kay, R. W.,** Survey of toxic hazards during vapour degreasing with trichloroethylene and 1,1,1-trichloroethane, *Ann. Occup. Hyg.,* 16, 417, 1973.
112. **Finney, D. J.,** *Probit Analysis,* 2nd ed., Cambridge University Press, London, 1964.
113. **Simaan, J. and Aviado, D. M.,** Hemodynamic effects of aerosol propellants. I. Cardiac depression in the dog, *Toxicology,* 5, 127, 1975.

Part II
Trichloroethylene

Chapter 6

INTRODUCTION TO TRICHLOROETHYLENE

Trichloroethylene, like methyl chloroform, is a solvent used in aerosol products. The preparation of the separate monographs for these two solvents which are both trichlorinated hydrocarbons allows us to compare the relative effects of each on the heart and lungs in identical animal systems and under identical conditions.

When methyl chloroform was introduced, its toxicity was demonstrated to be considerably lower than that of chloroform and carbon tetrachloride. There was limited study of its toxicity compared with trichloroethylene, even though this solvent was already widely used for the same purposes for which methyl chloroform was being proposed. Since trichloroethylene had already been used to administer general anesthesia, there was no reason to make a systematic study of the two trichlorinated hydrocarbons, one used medically and the other used industrially.

Recent events have compelled a reexamination of trichloroethylene. Consumer products containing these solvents are among those being abused by intentional inhalation. Furthermore, the solvents are unintentionally inhaled in the course of the normal use of aerosols and cleaning fluids, as well as in the industrial use of solvents.[1] There are reports of fatalities following industrial exposure and the abuse of products containing these solvents. However, there is no assurance that the inhalation of less than fatal amounts of the solvent is not hazardous to health.

This monograph on trichloroethylene has been prepared in the same manner as that of methyl chloroform (see Part I, Chapters 1 to 5). The literature on the medical use, toxicity, and pharmacology of trichloroethylene has been examined (see Chapter 7). Experiments involving inhalation of trichloroethylene have been completed (see Chapters 8 and 9). A comparison of the health hazards of trichloroethylene and methyl chloroform is included in the Concluding Remarks (see Chapter 10).

Chapter 7

REVIEW OF THE LITERATURE ON TRICHLOROETHYLENE

Trichloroethylene was discovered in 1864 by Fischer[2] in the course of preparing tetrachloroethane. The new volatile compound was subsequently used as a degreaser for machinery and as a solvent for organic products. The toxicity in the industrial use of trichloroethylene, consisting of four cases of trigeminal analgesia, was reported in 1915 by Plessner.[3] The literature on its toxicity was reviewed by von Oettingen[4,5] in 1937 and 1958, and by Browning[6] in 1953.

The first medical use of trichloroethylene was dictated by the analgesia encountered in poisoning, i.e., to treat trigeminal neuralgia.[7] The first use as a general anesthetic in man was by Striker et al.[8] in 1935, and the subsequent developments were reviewed by Atkinson[9] in 1960, and by Defalque[10] in 1961. The most recent review of the literature was written by Smith[11] in 1966. Subsequent to this date, there has been no concerted effort to summarize the literature, especially the recent publications which exceed two thousand articles. The publications which relate to the evaluation of inhalational toxicity of trichloroethylene have been selected and are discussed below.

A. Chemical and Physical Properties of Trichloroethylene

Synonyms: trichloroethene; ethinyltrichloride; the symbol TCE will be used in this monograph.

Trade names for anesthetic grade: Trilene; Triman (usually containing 1/10,000 thymol as a stabilizer and 1/200,000 waxoline blue to aid identification).

Trade names for industrial grade: Alk-Tri; Blacosolov; Ethyl Trichloroethylene; Ex-Tri; Hi-Tri; Neu-Tri; Nialk Trichlor MD; Nialk Trichlor MDA; Nialk Trichlor-Extraction; Nialk Trichlor-Technical; Nialk Trichlor X-1; Perm-A-Chlor; Perm-A-Chlor NA; Perm-A-Chlor NA-LR; Phillex; Stauffer Trichloroethylene; Triad-E; Trichlor Type 113; Trichlor Type 114; Trichlor Type 115; Trichlor Type 122; Triclene D; Triclene L; Triclean LS; Triclene MD; Triclene ME; Triclene R; Triclene; Triclene, High Alkalinity; Triclene, Paint Grade; Trichloroethylene-Degr. Gen. Solv.; Trichloroethylene Dual; Trichloroethylene Extraction Grade; Trichloroethylene High Purity; Tri-Paint Grade.

Structural formula

$$\begin{array}{ccc} Cl & & Cl \\ & C=C & \\ H & & Cl \end{array}$$

C = 18.28%; H = 0.77%; Cl = 80.95%

Molecular weight: 131.40.

Boiling range: 87.14–87.55°C; solidifies at −83°C.

Specific gravity: 1.465 at 20.4°C; viscosity: 0.55 centipoise.

Vapor pressure: 60 mmHg at 20°C.

Vapor density: 4.53 at 25°C.

Soluble: in ether, alcohol, and chloroform.

Solubility in water: 0.1 g/100 g water at 25°C.

Colorless, unsaturated aliphatic organic compound, mobile liquid at room temperature, with characteristic odor resembling that of chloroform.

Distribution coefficient:

Water/air	3 at 20°C; 1.6 at 37°C.
Blood/air	18 to 22 at 20°C; 8 to 10 at 37°C.
Plasma/air	16 to 20 at 20°C.
Fat/water	34.4.

Thermal stability: in the presence of oxygen and strong light or heat, carbonyl chloride (phosgene) and hydrochloric acid are formed, particularly at temperatures above 125°C. Thus, TCE should be stored in cans or dark glass bottles and should not be left in clear glass vaporizer containers for any prolonged period.[12] It has been suggested that the use of cautery in the operating room, particularly in the mouth, when TCE is the anesthetic agent, might give rise to dangerous amounts of phosgene.[13] TCE, as noted by Lehmann,[14] also decomposes in the presence of heat and alkali, first giving dichloroacetylene:

$$CHCl = CCl_2 + NaOH \rightarrow CCl \equiv CCl + NaCl + H_2O$$

Dichloroacetylene is itself highly toxic, giving rise to encephalitis in rodents, and it also decomposes further with hydrolysis, giving rise to phosgene, carbon monoxide, and various acids.[15]

Flammability: nonflammable in air; however, at temperatures above 25.5°C, the vapor will ignite if TCE is mixed with pure oxygen at concentrations between 10.3 and 64.5%.[16] Since a concentration of 10% TCE would never be used intentionally for anesthesia, it may be regarded as an agent of limited flammability.

Preparation of TCE: it is prepared by the chlorination of acetylene, yielding tetrachloroethane, which is then treated with lime slurry, giving TCE which is purified by distillation:

$$C_2H_2 + 2Cl_2 \rightarrow C_2H_2Cl_4$$
$$2C_2H_2Cl_4 + Ca(OH)_2 \rightarrow 2CCl_2CHCl + CaCl_2 + H_2O$$

Determination of concentration of TCE in gases and biologic fluids. The following methods are available:

Iodine pentoxide method – This method was described by Strayer[17] and depends upon passing the gas mixture over heated iodine pentoxide and asbestos. Iodine and hydrogen iodide are formed quantitatively and subsequently estimated by titration or colorimetry.

Fujiwara pyridine-alkali method – The determination requires quantitative extraction of TCE by the use of toluene and reaction with pyridine and potassium hydroxide. The color developed is read at 537 nm.[18,19]

Interferometry method – This method depends on the measurement of the change in refractive index of a gas produced by the presence of TCE vapor.[20]

Infrared spectrophotometry – This is a convenient and widely used method.[21-24] It estimates the vapor concentration by measurement of the selective absorption of radiation of wave length 11.8 μm which takes place when the radiation traverses TCE vapors.

Gas chromatography – In this method, the TCE is separated from the other components of the gas mixture by passage through a heated chromatographic column and is then estimated, usually by a katapherometer or hot wire detector, which measures the thermal conductivity of the vapor.[25]

B. Medical Uses

The circumstances surrounding the introduction of TCE as a general anesthetic have been reviewed by Atkinson,[9] Defalque,[10] Smith,[11] Dobkin,[26] and Parkhouse.[27] These review articles summarize its application as an inhalational anesthetic from 1935 to 1965. During the past decade, the modern inhalational anesthetics,[28] such as halothane, methoxyflurane, and fluroxene, have replaced TCE for various reasons such as extent of muscular relaxation and incidence of postoperative complication. A 1974 reevaluation of TCE by Pembleton[29] in 522 cases revealed that TCE is a satisfactory general anesthetic.

The initial use of TCE for the treatment of trigeminal neuralgia has been discarded because this inhalant is not specific for this disease. It is an effective analgesic in the following clinical situations: pregnant women early in labor,[30] during proctoscopic examinations,[31] for relief of postoperative pain,[32] for narcohypnosis,[33] and for angina pectoris.[34] The risks involved in the use of TCE as an anesthetic have not been completely evaluated. The technique of employing soda lime in the operating room to absorb exhaled carbon dioxide when TCE is used is dangerous because of the formation of dichloroacetylene which causes cranial nerve palsy.[35] There are known hazards of TCE on the fetus and heart which will be discussed below (see Section H), page 55.

C. Industrial Uses

The most comprehensive list of uses of TCE was compiled by Browning.[6] The chief industrial application is in the degreasing of metal parts in metalware factories, motor, machine, and railway works, and electrical works. The medical problems in workshops for degreasing have been reviewed.[36-38] The formation of dichloroacetyl chloride and phosgene from TCE in the atmosphere of welding shops has resulted in deaths from phosgene poisoning.[39-41]

TCE is also used in the boat, shoe, textile, and chemical industries, as well as in painting and enameling, photography, polishing of optical lenses, and extraction of caffein from coffee.

In dry-cleaning, TCE is one of the most widely used solvents. There have been reports of systemic poisoning among workers but it has not been possible to identify the causative agent since several chemicals are used in the same operation. Diagnosis by analysis of the urine for metabolites of TCE has been possible in a few cases of poisoning.[42-46]

D. Consumer Products

There is no available listing of consumer products containing TCE. A partial list of selected products has been made by Truhaut et al.[47] in France and by Gleason et al.[48] in the United States. In the latter, TCE is incorporated in cleansers for automobiles, buffing solution, spot remover, rug cleaner, disinfectant and deodorant, embalming dry shampoo, and cleanser for false eyelashes and for wigs. Two aerosol products include TCE and a fluorocarbon: a chimney sweep cleaner and a mildew preventive.

E. Human Investigation

A majority of the current information on the toxicology and pharmacology of TCE has been derived from observations made in human subjects who have been exposed occupationally, accidentally, and intentionally, the latter to induce anesthesia or as a form of abusive inhalation. The effects on the various organ systems are discussed in another section (see H, page 55). The observations that relate to the overall toxicity in man are discussed in the present section. The circumstances underlying the observations start from the uncertain situations such as occupational exposure in which the worker inhaled not only TCE but also other contaminants in the atmosphere. The list will conclude with the experimental inhalation in man of known concentrations of TCE so that it is possible to obtain definitive information on dose-responses.

1. Occupational Fatalities

It is generally assumed that if a worker dies from accidental exposure to a high level of TCE, the cause is the lethal amount of TCE inhaled.[49-51] The diagnosis is strengthened by isolating TCE from biological fluids and organs obtained postmortem. There is no information on the amount of TCE that is fatal to humans.

When there is no episode preceding death or disease, it is difficult to implicate chronic exposure to TCE. It is questionable if the level of exposure encountered ordinarily in work places can cause diseases involving the central nervous system, heart, liver, and kidneys.[52-54] It is also difficult to implicate TCE because of concomitant exposure to several different chemicals in the work environment. There are reports of exposure simultaneously to TCE and one of the following: perchloroethylene,[55] dichloroacetylene,[56] tetrachloroethane,[57] nitrobenzene,[58] gasoline,[59] naphthylamine,[60] and lead oxide.[61] The accompanying etiologic agent can include drugs taken by the patient,[62] stress,[63] and trauma.[64] The most vivid example of interaction is "degreaser's flush" which represents the dermal response to TCE and alcohol.[65] The vasodilation of the skin is caused by the combined action of both. There are also reports of neuropathy,[66] pulmonary edema,[67] and psychoses,[68] developing in workers exposed to TCE who have been drinking alcohol.

2. Addiction and Abuse

TCE is one of the solvents contained in cleansing fluid, glue, and aerosol products that are abused by teenagers. The most frequent lesion encountered during postmortem examination of addicts who indulge in sniffing of TCE are hepatic necrosis and nephropathy.[69-73] There are two cases of unusual lesions in the brain including ecchymotic foci in the dentate nucleus and cerebral vascular accident.[74,75] The reports of sudden death without pathological explanation indicate cardiac arrest as the probable cause.[76-79]

In the United States, the most commonly abused product containing TCE is cleaning fluid.[80,81] The experiences in other countries are different and are determined by availability of the products as well as reasons known only to the teenagers in England,[82] Belgium,[83] Scandinavia,[84,85] Italy,[86,87] Poland,[88-90] Czechoslovakia,[91,92] and Israel.[93,94] The incidence of sniffing has been increasing since 1950 as indicated by the growing literature on the subject. There is no reasonable estimate of the prevalence of this form of abuse.

3. Accidental Ingestion

Poisoning from ingestion of TCE is more readily diagnosed than that described above for abused inhalants and occupational exposure. The identity and quantity of ingested TCE can be verified by analysis of stomach contents, blood, or urine. The chemical identification has helped in the investigation of the criminal use of TCE as a poison,[95] and also of the wrong contents in a bottle labeled TCE.[96] The clinical picture of accidental ingestion includes signs of liver toxicity,[97-100] kidney malfunction,[101,102] and cardiac arrhythmias.[103,104] The late sequellae consist mostly of neurologic disturbances.[105,106] The mode of action of TCE in producing these lesions is discussed below (see Section H, page 55).

4. Hazards in an Operating Room

In spite of the repeated warnings not to use TCE in a closed circuit to administer anesthesia, there continue to be reports of accidental use in such a system.[107,108] There are also deaths from the use of TCE in therapeutic abortions; the incidence in a series of 200,000 women is 7 cardiac arrests.[109] The levels of TCE in operating rooms have been measured.[110,111] The concentration can be as high as 0.55 mg/l (103 ppm) directly over the expiratory valve and as low as 0.01 mg/l (1.9 ppm) in the periphery of the operating room. There were detectable levels of TCE in the end-expired air of the anesthesiologists from 2 to 7 hr following the administration of TCE to their patients.

5. Experimental Inhalation

All the published investigations on human volunteers consist of psychomotor and visual-motor performances. Stopps and McLaughlin[112] exposed one individual, twice, for an 8-hr period to each of four concentrations ranging from 0.535 mg/l (100 ppm) to 2.675 mg/l (500 ppm). There was no significant effect of the lowest concentration on psychomotor performance but there was a progressive decline in performance with 1.070 mg/l (200 ppm), 1.605 mg/l (300 ppm), and 2.65 mg/l (500 ppm).

Vernon and Ferguson[113] extended the range of concentration from 0.535 mg/l (100 ppm) to 5.35 mg/l (1000 ppm), to each of eight male subjects for 2 hr. Compared to control responses, the highest concentration had adverse effects on performance in the depth perception, steadiness, and pegboard tests. This had no effect on flicker fusion, form perception, or code substitution tests. All tests were unaltered by inhalation of 1.605 mg/l (300 ppm) and 5.35 mg/l (1000 ppm). The adverse effects of 5.35 mg/l were enhanced by the ingestion of alcohol.[114] This high concentration was reported to cause optokinetic nystagmus when administered for 2 hr to 12 subjects.[115]

Stewart et al.[116] conducted ten experiments in an exposure chamber. One experiment consisted of two subjects exposed for 4 hr to a mean level of 0.535 mg/l (100 ppm). There were no subjective effects. The remaining experiments consisted of exposing two or five subjects from 1 to 7 hr to a mean concentration of 1.07 mg/l (200 ppm). The untoward subjective responses were mild and inconsistently present. The only troublesome response noted was a sensation of mild fatigue and sleepiness in all five subjects during their fourth and fifth consecutive days of exposure to this concentration of TCE.

Salvini et al.[117,118] exposed six subjects for two 4-hr exposures, separated by a 1 1/2 hr interval, to a concentration of 0.589 mg/l (110 ppm). There was a significant decrease in performance ability in the perception test, the Wechsler memory scale, complex reaction time test, and manual dexterity test. They concluded that their results questioned the acceptability of a Threshold Limit Value of 100 ppm.

6. Absorption

The primary site of entry of TCE is in the lungs. In the experimental inhalation described above, there was a wide range of concentration of alveolar and venous blood levels of TCE in various subjects even though a fixed concentration was inhaled.[115] The wide range of levels in spite of a fixed concentration is determined by the same factors which influence the rate of absorption of a volatile anesthetic such as solubility coefficient, respiratory minute volume and cardiac output reviewed by Parkhouse[119] and by DuBois et al.[120] Kimmerle and Eben[121] exposed eight subjects to a concentration lower than that used by other investigators, i.e., 0.214 mg/l (40 ppm), for 4 hr, and four subjects to 0.267 mg/l (50 ppm) for 4 hr daily for 5 successive days. Following a single exposure, TCE could be detected in the blood for up to 4 days. However, contrary to expectation, repeated exposure did not cause a continuous increase of TCE level in the blood, indicating that the pharmacokinetic behavior of the solvent after acute exposure differed from that after subacute exposure. The accumulation of the metabolite, trichloroethanol, following repeated exposure suggested to the authors that the TLV of 100 ppm be reduced to 50 ppm.

For completeness, it should also be mentioned that TCE can be absorbed from the skin of hands that are soaked in the solvent.[122] The lungs are not only the primary site of absorption but also for excretion.

7. Blood Levels

Stewart et al.[123-125] exposed human subjects to 1.07 mg/l (200 ppm) for up to 3 hr. The blood showed a detectable level of TCE after half an hour and reached a peak of 6 ppm (1 mg/100 ml) in 2 hr. The solvent was detectable in the expired air for more than 5 hr after the exposure suggesting the usefulness of expired air for the evaluation of TCE exposure. Inhalation of anesthetic concentrations of 0.5% (5000 ppm or 26.75 mg/l) was accompanied by a rise in the brachial or radial arterial blood level to 6 or 7 mg/100 ml and a lower level (3 to 5 mg/100 ml) in the venous blood indicating that equilibrium is not attained for 2 or 3 hr.[126,127] The persistent arteriovenous difference in levels of TCE represents the uptake of the solvent in the tissues of the body.

8. Metabolism

The metabolism of TCE has been reviewed by Kelley[128] in 1974. There are three major metabolic transformations: (a) oxidation to chloral hydrate which takes place in the microsomal fraction of liver cells, (b) reduction to trichloroethanol, and (c) further oxidation of chloral hydrate to trichloroacetic acid. The metabolism of TCE in workers exposed to it has been examined by Sukhotina[129] and Sikhanova[130]. In human volunteers, there are sex differences in the nature of metabolites; trichloroacetic acid in females was found to be 2 to 3 times more than that in males and trichloroethanol was excreted twice as much in males as in females.[131] The concurrent administration of glucose and ethanol increased the amount of both metabolites.[132] The injection of disulfiram reduced the formation of trichloroethanol by 40 to 64% and of trichloroacetic acid by 72 to 87%, and an increase in excretion of the solvent by the lungs of up to 65%.[133] The inhibition of oxidation of TCE by disulfiram is a possible approach in the treatment of TCE intoxication since trichloroethanol is more toxic than the parent compound.

The accumulation of metabolites of TCE discussed in paragraph 6 above, has been further confirmed by recent observations on metabolism of TCE in human subjects undergoing experimental inhalation.[134,135] There is storage of both TCE and trichloroethanol in the tissues from which they are slowly released.

9. Urinary Excretion

After the experimental inhalation of TCE, the solvent is metabolized completely to the following: trichloroethanol that excretes rapidly in the urine, and trichloroacetic acid that accumulates in the body due to a high plasma protein binding rate. On the basis of total urinary excretion, the quantity of trichloroethanol and trichloroacetic acid is approximately double that of trichloroacetic acid.[136] The total amount excreted in the urine represents the quantity retained by the body and is about 60% of the TCE retained during experimental exposure.[137-139] There is a small fraction excreted in the feces, sweat, and saliva which is less than 10% of the total amount retained.

Measurement of urinary metabolites of TCE is used to evaluate occupational exposure in the United States,[140,141] Western Europe,[142-151] Eastern Europe,[152-154] and Japan.[155-158] In the last-mentioned country, Ikeda et al.[155] reported the excretion of metabolites in a patient addicted to TCE. After stopping the exposure, the urinary metabolites disappeared, and in three weeks, the psychotic symptoms cleared up. Nomiyama[156] has formulated equations to calculate the environmental concentration of TCE from either respiratory elimination or from urinary excretion. On the other hand, Stewart et al.[116] regarded urinary excretion as an unsatisfactory index of exposure and relied only on expiratory elimination. The difference between the two is that Nomiyama examined workers exposed to TCE for years and Stewart et al. studied volunteers who inhaled TCE for a few days. The results of subacute experimental inhalation do not deter from the long-accepted practice of using urinary excretion to estimate occupational exposure.

F. Metabolism and Disposition in Animals

Most of the animal experiments relating to metabolism of TCE do not consist of administering the solvent by inhalation. Radioactive carbon-labeled trichloroethylene was fed to rats and about 15% of the dose was excreted in the urine and the remainder was collected in the expired air.[159] The disposition of TCE was studied in rabbits following oral administration,[160] in guinea pigs after intraperitoneal injection,[161] and in dogs after oral ingestion.[162,163] The primary site of the metabolism of TCE is in the liver, which has been demonstrated by in vitro experiments in the rat,[164-167] rabbit,[168] and dog.[168]

The only investigations of metabolites involving inhalation of TCE in animals were performed in rats. Kimmerle and Eben[169] exposed rats 8 hr daily for 14 weeks to a concentration of 0.267 mg/l (50 ppm). Urinary excretion of trichloroethanol increased until the tenth week and then decreased slowly. In contrast, the amount of trichloroacetic acid excreted remained fairly constant. At the termination of the subacute exposure, there was enlargement of the liver as a consequence of enzyme induction but there were no pathological changes. Ikeda et al.[170,171] noted an increase in the formation of trichloroethanol in man when the concentration of TCE exceeded 0.267 mg/l (50 ppm). The results of inhalation experiments in rats confirm the observations in man that repeated exposure to TCE alters the nature of metabolism of this solvent which is probably due to an increase in hepatic enzymes (see also Section H, page 55).

G. Toxicologic Investigation in Animals

Thus far in this review, there has been no comparison between TCE and methyl chloroform because the information relating to the earlier sections does not exist. In the toxicologic investigation in animals, there are differences in potency of the two solvents, all of them indicating a higher level of toxicity for TCE compared to methyl chloroform.

1. Oral Administration

We have found only two published sources of oral lethal dosages of TCE in animals. Smyth et al.[172] reported that the LD_{50} for rats was 4.92 (3.75 to 6.46) g/kg. The lowest oral LD_{50} for dogs reported in the literature is 5.68 g/kg.[173] There are no comparisons of TCE and methyl chloroform performed by the same investigators. In a separate investigation, the oral LD_{50} for rats ranges from 10.3 to 14.3 g/kg methyl chloroform (see Chapter 2). The oral lethal ratio of methyl chloroform to TCE can be estimated to be 2.4 to 1.

2. Parenteral Injection

Klaassen and Plaa[174] conducted a comparison of both of the trichlorinated solvents in mice. The intraperitoneal LD_{50} for methyl chloroform and TCE suspended in corn oil, respectively, are 4.94 and 3.15 g/kg. Schumacher and Grandjean[175] estimated the LD_{50} for TCE as 1.25 g/kg (1.17 to 1.46). The intravenous lethal doses have also been reported in two animal species:[173] the LD_{50} for mice is 34 mg/kg and the lowest lethal dose for dogs is 150 mg/kg. The subcutaneous lethal dose for the rabbit is 1,800 mg/kg.[173]

3. Inhalation Toxicity: Single Exposure in Mice

The early investigations on inhalation toxicity were performed on mice. Lazarew,[176] in 1929, exposed mice for 2 hr and determined the concentration that produces narcosis (unconsciousness) and the lethal concentration of the solvent. His results are as follows:

	Narcotic concentration	Lethal concentration
TCE	25 mg/l	40–45 mg/l
methyl chloroform	45 mg/l	65 mg/l

From the above results, it can be noted that methyl chloroform is less toxic than TCE and that the inhalation lethal ratio is approximately 1.5 to 1. Furthermore, methyl chloroform is a weaker narcotic than TCE. The results of other investigators on TCE are as follows:

Cresutelli[177] (1933)	17.6 mg/l for 2 1/2 hr exposure; death 1 hr after exposure.
	46.9 mg/l for 1.4 hr; death 110 min after exposure.
	50.0 mg/l for 1.2 hr; death 67 min after exposure.
Gehring[178] (1968)	30 mg/l; anesthetic time$_{50}$ = 46 min and lethal time = 585 min

The investigation of Gehring[178] is of special significance because he confirmed essentially the results of Lazarew. They are as follows:

	Lethal concentration 6 hr	Anaesthetic time at lethal concentration
TCE	29 mg/l (5500 ppm)	46 min
methyl chloroform	73 mg/l (13,500 ppm)	16 min

The inhalation lethal ratio is 2.5 to 1 and the margin of safety for the anesthetic effect of methyl chloroform is wider than that of TCE.

4. Inhalation Toxicity: Single and Repeated Exposures in Rats

Carpenter et al.[179] exposed rats for 4 hr and reported lethality in half of the animals at the following concentrations: 43.43 mg/l (8,000 ppm) for methyl chloroform and 10.7 mg/l (2,000 ppm) for TCE. Their results are based on 6 rats exposed to one concentration. Siegel et al.[180] used 16 rats each at three levels of concentration and reported the lethal concentration$_{50}$ for 4 hr as follows: TCE = 66.9 mg/l (12,500 ppm) and methyl chloroform = 100 mg/l (18,400). The lethal ratio for both is 1.5 to 1.

The first investigation that includes both single and repeated inhalational toxicity of TCE was reported in 1951 by Adams et al.[181] The least severe single exposure causing death of 100% of the rats was as follows: 107 mg/l (20,000 ppm) for 0.4 hr; 64.2 mg/l (12,000 ppm) for 1.4 hr and 13.5 mg/l (3,000 ppm) for 7.0 hr. For each of the three concentrations, there was no depression of the central nervous system when exposure lasted for 0.3 hr, 0.6 hr and 1.4 hr, respectively. Rats were also exposed 7 hr daily, 5 days a week, for 6 months. The exposure to 1.07 mg/l (200 ppm) did not result in any detectable injury. Exposure to 2.15 mg/l (400 ppm) depressed the growth of male rats and increased the liver and kidney weight.

The second reported subacute inhalation of TCE was done by Prendergast et al.[182] They exposed 15 rats, 8 hr daily, 5 days a week for 6 weeks at a concentration of 3.8 mg/l (700 ppm). There were no deaths and the only sign observed was nasal discharge which was also seen in the control rats. Lung congestion was noted in an occasional rat. There was no evidence of chemically induced histopathologic changes in the visceral organs of rats. Continuous exposure for 90 days to 0.189 mg/l (36 ppm) did not cause any visible sign of toxicity.

5. Inhalational Toxicity: Other Animal Species

McCord[183] reported the following concentrations that were lethal to rabbits: 26.75 mg/l (5,000 ppm) for 14.28 hr, 53.5 mg/l (10,000 ppm) for 2½ hr, and 107 mg/l (20,000 ppm) for 2 hr. Matruchot[184] exposed guinea pigs at a concentration of 140 to 160 mg/l (26,180 to 29,290 ppm) to determine the duration until the animals died. His results were expressed in terms of relative toxicity; methyl chloroform was 5 times less toxic than TCE. Adams et al.[181] exposed guinea pigs, rabbits, and monkeys 7 hr daily, 5 days a week for about 6 months. The maximum concentrations tolerated were as follows: guinea pig 0.535 mg/l (100 ppm), rabbit 1.07 mg/l (200 ppm), and monkey 2.14 mg/l (400 ppm).

Prendergast et al.[182] exposed guinea pigs, dogs, rabbits, and monkeys at the same concentrations used for rats (see above). Repeated exposure to 3.8 mg/l (700 ppm) for 15 days and to 0.189 mg/l (36 ppm) for 90 days did not cause any visible sign of toxicity.

H. Pharmacodynamics

Thus far in this review, the information derived from animal investigations has been separated from that derived from human investigations. This section combines both in an attempt to define the toxicity of TCE for each organ system.

1. Neurotoxicity

The use of TCE as a general anesthetic and as an analgesic is based on its narcotic properties. The symptoms of dizziness and loss of consciousness in acute poisoning can be readily explained by the anesthetic action of TCE. There are other neurologic manifestations in workers exposed to TCE that include mental disorders,[184-186] cranial neuropathy,[187-192] peripheral neuropathy,[193-195] and atrophic lesions of the brain[196,197] and spinal cord.[198] These lesions have not been reproduced in animals inhaling TCE so that there is no information on the mechanism by which TCE produces neurologic disease. Parenteral injection of TCE (18 to 133 g total dose) in rabbits[199] in a trial lasting for 29 days, resulted in nerve cell damage similar to that seen in human poisoning.

The alterations in the electroencephalogram in patients undergoing general anesthesia with TCE[200] are manifestations of its depressant action on neurons. The most conspicuous sign is exaggeration and increase in the frequency of alpha waves, an effect seen also in the electroencephalogram of workers exposed to TCE.[201-210] The electroencephalogram and cortical potentials in animals[211-217] show a depression of activity due to an effect on the synapses by TCE, similar to that elicited by other general anesthetics. The brains of mice exposed to TCE show a reduction in high-energy phosphates[218] and inhibition of a brain protease[219] which are manifestations of depression of neuronal activity.

Behavioral studies in rats have been helpful in detecting the threshold dose of TCE. Grandjean[220-225] exposed rats daily for 44 weeks to concentrations ranging from 0.107 mg/l (20 ppm) up to 4.28 mg/l (800 ppm). The results varied with the type of testing; there was no modification of either the conditioned responses or the response time of rats who had been trained to climb a rope in order to reach a feeding trough where they found a reward. On the other hand, the spontaneous climbs were more frequent. Alternating left to right turning behavior was significantly increased at 2.14 to 3.21 mg/l (400 to 600 ppm); swimming speed in water was increasingly reduced by exposure to 3.2 mg/l (400 ppm).

Most other behavioral studies have been performed in rats[226-228] and guinea pigs[229-231] exposed to 1.07 mg/l (200 ppm) or higher concentrations of TCE. The only exception was the study of Horvath and Formanek[232] who used 0.400 mg/l (76 ppm). After 72 to 121 days of exposure, there was a disturbance of the dynamics of cortical activity. Only four rats were used and it will be necessary to increase the number of animals before making a definitive statement that the threshold for behavioral changes is as low as 0.400 mg/l (76 ppm). The analgesic effect of TCE inhalation has been more readily demonstrated in mice[233] and rabbits[234] than in human subjects, with pain induced by ischemia of the extremities.[235] Unfortunately, the effective analgesic dose was not correlated with the anesthetic dose and the lethal dose so that it is not possible to rate the safety factor of TCE in animals.

2. Cardiotoxicity

The most frequent manifestation of acute poisoning from TCE is cardiac arrhythmia in humans. The most direct proof that TCE can cause ventricular extrasystoles, fibrillation, and cardiac arrest is that these changes can be demonstrated in the electrocardiogram of subjects who have accidentally ingested TCE.[236-238] The appearance of these abnormalities in workers exposed to TCE, however, makes it more difficult to establish TCE as the causative factor.[239-242] There are changes in cardiac function that are seen in a majority of workers exposed to TCE which include the following: increase in cardiac output,[243] duration of isometric contraction, and duration of the period of tension.[244] When exposure to the solvent was suspended, these signs of cardiac malfunction disappeared indicating that TCE was the cause.

The experiments in animals have consistently demonstrated a depression of myocardial contractility. In 1921, Kiessling[245] observed depression in the perfused frog heart by TCE with a potency that was 6.2 times less than that of chloroform. In the isolated guinea pig heart, Bianchi et al.[246] and Matturro[247] observed the following effects: a concentration of 0.53% in the perfusate caused a transitory arrhythmia without an effect on contractility; 1.06% caused a reduction in the force of contraction, arrhythmia, and reduction in coronary flow.

Arrhythmia can be provoked by the injection of sympathomimetic drugs in the dog inhaling 0.5 or 1.0% TCE in air.[248] This form of arrhythmia can be prevented or terminated by the administration of a beta-blocking agent in the dog[249-251] and rabbit.[252,253] Hypoxia[254-256] and hemorrhage[257] also sensitize the heart to arrhythmia in the course of anesthesia by TCE. The heart is capable of responding to hypoxia and hyperoxia by increasing or decreasing cardiac output in dogs that are anesthetized with TCE.[258] The end result is a heart with low cardiac output but still capable of responding to sympathetic influences.

3. Vasotoxicity

The blood vessels show dilatation as well as constriction, depending upon the organ, when there is an elevation of content of TCE in the blood. Ellis[259] observed that patients who underwent surgery for the correction of convergent strabismus lost significantly more blood when anesthetized with TCE, as compared to being anesthetized with halothane. The vasodilatation in the ocular cavity also includes the cerebral vessels which show an increase in blood flow accompanied by elevation of cerebrospinal fluid pressure.[260-264] It is possible that these changes seen in acute inhalation of TCE may occur during occupational exposure and relate to the occurrence of cerebral hemorrhage seen in some workers.[265]

The blood vessels of the extremities show an increase in blood flow if not diseased,[266] or a decrease if ischemic.[267] The latter is due to the shifting of blood from the areas supplied with abnormal arteries to those with a normal supply and which dilate in response to TCE. There are no measurements of blood flow in the other special vascular beds. The end result is expected to be a combination of the local vasodilator action of TCE and the reflex vasoconstriction from a fall in blood pressure if the solvent is present in concentrations high enough to reduce cardiac output and reduce the arterial blood pressure.[268-271]

4. Pneumotoxicity

Congestive atelectasis has been reported to occur in workers exposed to TCE[272] and in patients who accidentally ingest TCE.[273] In its use as a general anesthetic, TCE has a bacteriostatic effect[274-276] so that this solvent does not increase the susceptibility of the lungs to bronchopulmonary pathogens. The respiratory depression seen in patients under TCE anesthesia[277-281] would reduce tidal volume which would promote atelectasis since TCE also causes bronchoconstriction[246] and interference with the cough reflex. In animal experiments, the effects of TCE on respiration are brought about by a combination of central depression[282-284] and stimulation and even paralysis of receptors in the lungs.[285-287] The inhalation of TCE reduces the vasoconstrictor action of hypoxia in the cat lung,[288] but does not protect the guinea pig from bronchoconstriction of anaphylaxis.[289] Most other inhalational anesthetics have the same interactions. There is a quantitative difference in the safety index among the anesthetics. TCE has the narrowest margin between anesthetic dose and respiratory arrest in mice and rats compared to either chloroform or halothane.[290,291] Truhaut et al.[292] explained the respiratory arrest as a manifestation of central nervous system depression, an effect that unsaturated compounds, such as TCE, exert more than saturated compounds, such as methyl chloroform.

5. Hepatotoxicity

Hepatic necrosis occurs in man following TCE intoxication, either by inhalation or by ingestion.[293-295] Occupational exposure, to TCE if prolonged for years, can cause an elevation of serum transaminases indicating damage to the liver parenchyma.[296-301] Brief inhalation in the course of administering general anesthesia does not influence liver function.[302,303]

Experimental exposure of animals has led to a determination of the comparative hepatoxicity of TCE and methyl chloroform. Klaassen and Plaa[174] administered the solvents by intraperitoneal injection in mice and reported the following effective $dose_{50}$ that would cause elevation of serum glutamic-pyruvic transaminase activity: 25 mM/kg of methyl chloroform and 18 mM/kg of TCE. The hepatoxicity ratio of both is 1.3 to 1. Gehring[178] administered the solvents by inhalation for several hours and reported the following ED_{50} in mice: methyl chloroform 73.3 mg/l (13,500 ppm) for 595 min, and TCE 29.4 mg/l (5,500 ppm) for 400 min. The margin of safety between anesthetic dose and hepatotoxic dose of TCE was narrower than that for methyl chloroform. A comparison of TCE with other solvents in mice has been made by other investigators. Kylin et al.[304,305] and Ikeda et al.[306] observed that TCE was less hepatotoxic than its tetrachloro analogue. The metabolites of TCE, as well as the accompanying biochemical changes in the liver, have been elucidated in mice,[307,308] rabbits,[309,310] dogs,[311] and rats.[312-321]

The biochemical changes in the liver brought about by TCE are similar to those by other inhalational anesthetics.[322,323] The investigation of the influence of pretreatment with phenobarbital on the hepatotoxicity of TCE has led to conflicting results. Cornish et al.[324] failed to find an enhancement of hepatotoxicity in rats when TCE was administered intraperitoneally. On the other hand, Carlson[325] observed an enhancement of hepatotoxicity when TCE was administered by inhalation for 2 hr. Pretreatment with phenobarbital is known to produce hepatic enzymes which accelerate the metabolism and increase the toxicity of metabolites of TCE, only when inhaled.[325] Although there is no enhancement exerted on injected TCE by phenobarbital, it has been possible to demonstrate increased hepatotoxicity when TCE is injected with isopropyl alcohol or acetone.[326] The potentiation of TCE toxicity by ethanol has been observed in rats,[327] rabbits,[328] and man.[329,330] The mechanism for the interaction has not been elucidated. Blood alcohol levels are not enhanced by exposure to TCE.[331]

6. Nephrotoxicity

Prior to 1959, there were no reports of renal failure caused by inhalation of TCE. Since that year, nine cases have appeared in the literature.[332-336] Experimental inhalation of TCE has been demonstrated to cause nephropathy in rats[337] and rabbits.[338,339] In dogs, there was no effect of intraperitoneally injected TCE on the excretion of phenosulfonphthalein.[311] Clearance techniques have not been applied to animals that have been exposed to TCE via the respiratory tract.

Dermatotoxicity

Exposure to TCE has been reported to cause generalized dermatitis, both of the exposed and nonexposed areas of the skin.[340-343] It has not been possible to distinguish how much of the reaction is a hypersensitivity response. The mucous membranes of the eyes, nose, and mouth are also influenced.

8. Embryo- and Fetotoxicity

Pregnant mice and rats were exposed to the following levels of TCE: 1.605 mg/l (300 ppm)[344] and 0.34 mg/l (65 ppm).[345] There were no adverse effects on embryonal or fetal development. However, this does not mean that TCE is free of hazard in pregnancy. This solvent has been demonstrated to cross the placental barrier[346] and may be responsible for certain of the respiratory difficulties of the newborn.[347-353] During labor, TCE reduces uterine motility[354-357] and its use as an analgesic has been accompanied by an increase in maternal and fetal mortality.[358,359] Newborn rats exposed continually to 0.004 mg/l (1 ppm) have been reported to show retardation in growth.[360] There has been no confirmation of this extraordinary effect of TCE.

9. Other Organ Systems

There are scattered reports that toxicity to TCE is manifested by lesions appearing in the adrenal glands,[361,362] musculoskeletal system,[363,364] ophthalmic system,[365-367] digestive system,[368-372] hematopoietic system,[373-379] fluid and electrolyte balance,[380,381] and immunologic system.[382-386] The pattern of toxicity is so varied and the occurrence is so rare that it is not possible to make general statements as to whether or not these organs are involved in occupational exposure to TCE.

1. Clinical Toxicology

The primary manifestations of acute poisoning from TCE relate to depression of the central nervous system. The subject initially complains of dizziness, nausea, vomiting, sleepiness and then lapses into a coma. It is estimated that the minimal concentration that would cause loss of consciousness in man is 16.05 mg/l (3,000 ppm).[387] The corresponding ingested dose that would lead to coma ranges from 50 to 150 ml in an average adult. Further depression of the central nervous system would lead to respiratory arrest and death.

Most victims of acute poisoning, when seen by physicians, die from primary cardiac arrest with secondary respiratory failure. If the patient survives the cardiac episode, then he is likely to manifest disturbances in function of the lungs, liver, and kidneys which, if severe enough, would cause death.[388-394]

Occupational exposure to TCE, if severe enough, may be accompanied by diseases of the central nervous system, heart, lungs, liver, and kidneys.[395-398] The nature of these diseases has been reviewed in Sections E and H. In some instances, it is impossible to state whether the disease is related to occupational exposure or is independent of it.

The exposure to TCE can be readily recognized by analysis of the expired air, blood, and urine.[399-402] The detection of metabolites in the urine is a simple test that can be readily performed in a clinical laboratory.

The treatment of acute poisoning is largely supportive in nature and includes administration of fluids, artificial respiration, and hemodialysis.[403-408] Sympathomimetic drugs should be avoided because of the sensitization of the heart to arrhythmia. The following drugs have been tested in animals and patients with encouraging results: propranolol to treat tachyarrhythmias,[409] atropine against bradyarrhythmias, ethinamate to stimulate respiration,[410] mineral oil to retard gastrointestinal absorption,[411] and disulfiram to reduce the metabolism of TCE to trichloroethanol, the latter being more toxic to the liver and heart than the parent compound.

J. Threshold Limit Value

In the United States, the American Conference of Government Industrial Hygienists[412] had originally set the Threshold Limit Value (TLV) for TCE at 1.094 mg/l (200 ppm). However, in 1961 the TLV was reduced to 0.547 mg/l (100 ppm).[413] This new level was carried through in 1963[414] and in 1971.[415] The Occupational Safety and Health Administration is currently in the process of approving the Criteria Document for TCE[1] which recommends that a TLV expressed as a time-weighted average exposure for an 8-hr workday continue at 0.547 mg/l (100 ppm), and that the present ceiling be reduced from 1.07 mg/l (200 ppm) to 0.80 mg/l (150 ppm).[416,417]

The continuation of the TLV at 0.547 mg/l (100 ppm) in the United States has been commented upon by Messite[418] who pointed out the urgent need for more epidemiological data based on controlled studies of TCE exposures in industry. In several European countries, the Maximum Allowable Concentration (MAC) has been set at 0.273 mg/l (50 ppm) or even lower. The reasons for a conservative level are outlined in review articles[419-428] and have also been stated in various parts of this review.

K. Conversion Table for Concentrations of Trichloroethylene

This review of the literature concludes with a table that is useful in converting concentrations of TCE on a volume/volume basis to weight/volume. The conversion table is based on the following constants for TCE: molecular weight of 131.40; volume occupied by one molecular weight of TCE at 26.3°C = 24.56 liters; therefore, 5.350 g of TCE produces 1 liter of gas.

mg/l	ppm	v/v%
0.00535	1	–
0.0535	10	0.001
0.1	19	–
0.535	100	0.01
1.0	187	–
5.35	1000	0.1
10.	1869	–
53.50	10,000	1.0
100.	18,692	–
267.5	50,000	5.0
500.	93,460	–
535.0	100,000	10.0

Chapter 8

ACUTE INHALATIONAL, ORAL, AND INTRAPERITONEAL TOXICITY OF TRICHLOROETHYLENE IN MICE

In the course of investigating the toxicity of methyl chloroform, it became apparent that the usual practice of estimation of LD_{50} could be improved by recording the electrocardiogram (see Part I, Chapter 3). More specifically, there was a dose-dependent decrease in heart rate and in QRS potential and a prolongation of the PR interval and QRS duration. It was concluded that lethality was correlated with these electrocardiographic changes brought about by methyl chloroform administered by the inhalational, oral, and intraperitoneal route. The present study on trichloroethylene (TCE) was designed to compare its electrocardiographic effects with those of methyl chloroform.

A. Methods

Experiments were carried out on male mice of the Swiss-Webster strain, with weights ranging from 20 to 25 g, divided into three series according to the route of administration of TCE. Preliminary experiments were performed to determine approximately the dose that kills all mice, and five lower doses were then administered to determine the LD_{50}.

Series I: oral route – Mice were divided into five groups, each of 10 mice, to which the following doses of TCE were administered: 2, 3, 4, 5, and 6 g/kg.

Series II: intraperitoneal injection – To each group, which consisted of 10 mice, a varying dose of TCE was administered (0.5, 1.0, 2.0, 3.0, and 4.0 g/kg).

Series III: inhalational route – Four groups, with 10 mice in each, were exposed to 0.054, 0.268, 0.535, and 1.070 g/l (1.0, 5.0, 10, and 20%) concentrations of TCE. Electrocardiographic changes (lead II) were monitored for 5 mice from each group, selected at random during 5 min of exposure and at repeated intervals up to a period of 24 hr after exposure.

Since it was not possible to calculate the 5-min LD_{50} because of shortness of exposure time, another five groups with 10 mice in each were exposed for 20 min to 0.054, 0.162, 0.268, 0.535, and 1.070 g/l (1.0, 3.0, 5.0, 10, and 20%) of TCE. The animals were exposed to the gas in a chamber into which the appropriate gaseous mixture flowed at a rate of 3 liters/min.

Various concentrations were prepared by volatizing into a known volume of air, a volume of liquid TCE, calculated to give the desired concentration at standard pressure and 25°C. The process of volatilization was aided by passing air into the container which was immersed in hot water.

Calculation of the lethal $dose_{50}$ – Survival was observed in all series for a period of 24 hr following administration of TCE. The LD_{50} was calculated according to the probit method. Electrocardiographic data were analyzed by paired comparisons, the criterion for significance being P less than 0.01.

B. Results

Oral lethal $dose_{50}$ – Mice in this series received TCE orally. Results are depicted in Table 8.1, which shows that doses of 2, 3, 4, 5, and 6 g/kg were associated, respectively, with mortality rates of 30, 60, 60, 80, and 100%, counted 24 hr later. The calculated LD_{50} (95% fiducial limits) was 2.85 ± 0.55 g/kg.

Intraperitoneal lethal $dose_{50}$ – Animals in this series were injected with TCE intraperitoneally. Results are summarized in Table 8.2 which shows that TCE in a dose of 0.5 g/kg was nonlethal. The injections of doses of 1, 2, and 3 g/kg were associated with mortality rates of 40, 80, and 100%, respectively. The calculated LD_{50} (95% fiducial limits) was 1.20 ± 0.31 g/kg within 24 hr after administration.

Inhalational lethal $concentration_{50}$ – This series comprised mice that were exposed to varying concentrations of TCE for 5 min. It was not possible to determine the LC_{50} since the observed mortality rate did not seem to correlate with the concentration of TCE. For example, the four ascending concentrations (0.054, 0.268, 0.535 and 1.070 g/l) caused death rates of 0, 20, 30 and 20%, respectively. Because of difficulties encountered in preparing high concentrations of TCE, and the nonlinearity of animal response, it was decided to prolong the exposure time to 20

TABLE 8.1

The Effect of Various Doses of Trichloroethylene, Administered Orally on the Survival of Mice

Groups	Dose (g/kg)	% Death	LD_{50} (g/kg) 24 hr	Regression coefficient
1	2	30		
2	3	60		
3	4	60	2.85 ± 0.55	0.9701
4	5	80		
5	6	100		

TABLE 8.2

The Effect of Various Doses of Trichloroethylene, Injected Intraperitoneally, on the Survival of Mice

Groups	Dose (g/kg)	% Death	LD_{50} (g/kg) 24 hr	Regression coefficient
1	0.5	0		
2	1	40		
3	2	80	1.20	
4	3 ·	100	±	0.9819
5	4	100	0.31	

TABLE 8.3

The Effect of Various Concentrations of Trichloroethylene Administered by Inhalation for 20 min on the Survival of Mice

Groups	Calculated vapor concentration % v/v	g/l	% Death	LC_{50} (g/l) 24 hr	Regression coefficient
1	1	0.054	0		
2	3	0.162	30		
3	5	0.268	30	0.220±	
4	10	0.535	90	0.099	0.8895
5	20	1.070	100		

min. Thus, five groups of mice, with 10 animals in each, were exposed for 20 min to the following concentrations of TCE: 0.054, 0.162, 0.268, 0.535, and 1.070 g/l. Results are summarized in Table 8.3 which shows that no death was produced by the lowest concentration. The mortality rate increased with an elevation in the vapor concentration of TCE, reaching 100% with 1.070 g/l. The LC_{50} was 0.220 ± 0.099 g/l.

Electrocardiographic changes – The following four parameters were measured from the electrocardiogram: heart rate (beats/min), PR interval (msec), QRS duration (msec), and QRS potential (mV). There were no significant changes in any of the above-mentioned parameters at any concentration save in a few scattered instances. Data on the mice that died at various concentrations were pooled together and the features of the electrocardiograms are summarized in Table 8.4. The most frequent alterations in the electrocardiogram were slowing of the heart rate, widening of the PR interval, and prolongation and depression of the QRS potential.

Among the surviving mice, only bradycardia appeared in more than half (in 55%). This was the only electrocardiographic change that appeared in both surviving and nonsurviving mice and can be regarded as the most consistent response to inhalation of TCE, regardless of the concentration. Among the surviving mice, there was prolongation of the PR interval (in 42%), and shortening occurred rarely (in 5.3%). The QRS complex responded in a variable fashion either increased, decreased, or not altered at all (Table 8.5).

C. Discussion

The electrocardiographic effects of inhalation of TCE were different from those reported for methyl chloroform. With the latter, there was a dose-dependent decrease in heart rate and in height of QRS potential and an increase in the PR interval and in the duration of QRS potential. With the inhalation of TCE, there was no consistent pattern in the electrocardiogram that was dose related. The mice that died showed bradycardia and widening of the PR interval, changes similar to those seen in human cases of poisoning (see Chapter 7).

The difficulties in obtaining the lethal inhalational dose from a 5-min exposure to TCE appear to be related to the respiratory response. The unanesthetized mouse immediately reacts by reflex depression of breathing from irritation of the upper respiratory tract. The uptake of TCE in the alveoli becomes unpredictable, especially because the initial amount absorbed would further depress the respiratory center. During the period of inhalation lasting for 5 min, the amount of solvent absorbed would be so variable that a dose-related mortality could not be demonstrated.

The 20-min inhalation resulted in a dose-related mortality, with the estimated LC_{50} of 220 mg/l

TABLE 8.4

Percent Changes in Different Electrocardiographic Parameters Calculated from Data of 6 Mice that Died During or Immediately After Exposure to 0.268 to 1.070 g/l Concentrations of Trichloroethylene for 5 min

	Percent changes: mean ± SE				
Parameter	1 min	2 min	3 min	4 min	5 min
Heart rate	+5.1 ±14.8	−21.0 ±24.5	−38.1 ±19.67	−33.2 ±28.99	−34.4 ±29.37
PR interval	+17.6	+8.8	+20.6	+17.6	+5.9
QRS duration	+22.8 ±16.91	+20.4 ±17.43	+15.9 ±20.4	+2.2 ±12.86	+27.6 ±15.38
QRS potential	+8.9 ±20.37	−24.8 ±14.08	−24.0 ±10.7	−20.2 ±13.60	−40.1 ±14.4

TABLE 8.5

Percent of Mice that Survived Showing Significant Increase or Decrease or Insignificant Change in the Electrocardiogram – Data Pooled from 24 Mice Exposed to Concentrations of from 0.268 to 1.070 g/l of Trichloroethylene

Parameter	Significant increase	Significant decrease	Insignificant change
Heart rate	25%	55%	20%
PR interval	42%	5.3%	52.6%
QRS duration	8.3%	12.5%	79.2%
QRS potential	37.5%	41.7%	20.8%

(41,000 ppm). This is the first report of inhalational dosage of less than 1 hr in mice. In all previous studies, periods of 2 hr or longer have been used.[176-178]

The intraperitoneal LD_{50} in the above experiments is 1.2 g/kg, which is similar to 1.25 g/kg reported by Schumacher and Grandjean.[175] Klaassen and Plaa[174] derived a larger value of 3.15, using corn oil as a suspending agent, which retards absorption from the peritoneal cavity.

The oral LD_{50} of 2.85 g/kg reported above has not been hitherto employed. The value is 2.4 times larger than the intraperitoneal lethal dose, a relationship which suggests good absorption of TCE in the gastrointestinal tract.

D. Summary

The LD_{50} and LC_{50} of TCE determined in male mice are as follows: oral 2.85 g/kg, intraperitoneal 1.20 g/kg, 20-min inhalation 0.220 g/l. There was no correlation between electrocardiographic changes and lethality, indicating that death was caused primarily by depression of the central nervous system and respiratory arrest.

Chapter 9

ACUTE INHALATION TOXICITY OF TRICHLOROETHYLENE IN DOGS

The experiments in mice reported in Chapter 8 were informative in characterizing the electrocardiographic effects of the inhalation of TCE. However, it was not possible to establish a dose-response relationship because of the primary effects of TCE on respiration. Upon starting the inhalation, there was depression and irregularity of breathing which influenced the uptake of TCE, and the unpredictability of the respiratory response caused the nonuniformity in the electrocardiographic changes.

The next animal species selected was the dog, and the experiments reported below were designed to examine the respiratory, bronchopulmonary, and cardiovascular effects of the inhalation of TCE. As a general anesthetic, TCE in sufficient concentration is known to depress respiration. The bronchopulmonary effects are not known. The cardiovascular actions consist of a depression of contractility of the isolated heart which has been known for some time,[245-247] but there is no measurement of maximal rate of rise of left ventricular pressure (dp/dt) in the intact heart. While a reduction in cardiac output has been noted in man and in the dog under general anesthesia,[254-258] the effect of subanesthetic concentrations has not been identified. The techniques for simultaneously measuring cardiac function, systemic vascular resistance, and pulmonary vascular resistance, developed for the study of methyl chloroform (see Chapter 4), have been applied to the investigation of trichloroethylene (TCE).

A. Methods

Closed chest – Experiments were carried out on mongrel dogs of either sex weighing between 14 and 23 kg with an average of 18.2 kg. The dogs were injected with 1 mg/kg morphine sulfate subcutaneously and half an hour later they were anesthetized by intravenous injection of 70 mg/kg chloralose. The trachea was then exposed and two cannulas, one directed toward the lung and the other toward the nose, were then inserted into the trachea. By means of this technique it was possible to administer TCE either to the lungs, bypassing the nose, pharynx, and larynx, or to the upper respiratory tract separately from the lower. Arterial blood pressure was measured by introducing a catheter through the carotid artery, attached to a P23AA Statham pressure transducer. The tracheal cannula was connected to a mesh screen Fleisch pneumotachograph and a differential pressure transducer model 270 was used to measure the pressure difference across the screen. Meanwhile the signal corresponding to air flow was integrated and recorded as tidal volume. Another differential transducer (Sanborn, model 267 B) was used to measure the pressure difference between the trachea and the intrapleural space. Pulmonary resistance and compliance were estimated from measurements of tracheal air flow and transpulmonary pressure plotted on an X, Y plotter. Recordings were made on a six-channel Sanborn 7700 recorder. After allowing the preparation to stabilize for 30 min, the following concentrations of TCE, volatilized in air, were inhaled via the lower tracheal cannula or were forced through the upper cannula toward the nose: 0.1% (5.35 mg/l), 0.5% (26.7 mg/l), 1.0% (53.5 mg/l), and 5.0% (267.5 mg/l). The administration of each concentration, lasted for 10 min alternating with 10 min of exposure to room air. The administration of TCE was repeated after cutting the vagi in the neck region.

Open chest – Experiments were carried out on adult mongrel dogs of either sex, weighing between 21 and 34 kg with an average of 26 kg. After injection of pentobarbital sodium (30 to 35 mg/kg i.v.) all the dogs were artifically ventilated with room air via an endotracheal tube, using the Starling Ideal respirator. The chest was opened on the left side at the fourth intercostal space. Noncannulating flow probes with an internal diameter of 12 to 16 mm were placed around the pulmonary artery to measure the pulmonary arterial blood flow. Flow probes were connected to a Statham SP 2202 electromagnetic flowmeter. The left atrial pressure was measured by introducing a catheter through a small cut in the left atrium, attached to P23AA Statham pressure transducer. Pulmonary arterial blood pressure was measured by introducing a catheter into the pulmonary artery to the left lower lobe. Thoracotomy was then carried out in the sixth intercostal space and the left ventricular pressure was measured by inserting a catheter tip into the apex of the heart. The signal was fed to a differentiating circuit, by the use of a derivative computer Type 8814A, and the left ventricular maximal rate of rise of pressure was recorded. The right carotid artery was then cannulated to measure the aortic pressure. All pressures were measured by the Statham P233A pressure transducer and all responses were recorded on a six-channel Sanborn 7700 recorder. After allowing the preparation to stabilize for 30 min, TCE vapors in air were administered through the inlet of the pump for 5 min. The following progressively increasing concentrations of TCE were administered: 0.01% (0.535 mg/l), 0.05% (2.675 mg/l), 0.1% (5.35 mg/l), 0.25% (13.37 mg/l), 0.5% (26.7 mg/l), and 1% (53.5 mg/l). About 10 min was allowed to elapse before a subsequent concentration was inhaled.

To find out any interaction between TCE and trichlorofluoromethane (FC 11) the latter was administered in a concentration of 0.5% that was found to bring about minimal effects in a similar preparation for 5 min. After recovery, a mixture of 0.5% FC 11 and 0.05% TCE was then administered. The sequence of administration of FC 11 and the mixture of both was alternated. All gaseous mixtures were prepared by volatilizing a known volume of TCE and/or FC 11, calculated to give the appropriate volume of gaseous phase at standard pressure and 25°C into the calculated volume of air measured by flowmeter (Fischer and Porter Co., Model 10 A 3565). Gas chromatography of the mixture revealed the concentration of TCE to within 5% of the calculated value. Data were analyzed by paired comparisons, and student-t-test, the criterion for significance being P less than 0.05.

B. Calculations and Abbreviations of Hemodynamic Measurements

MPAP: Mean pulmonary arterial pressure in cm H_2O, measured from the pulmonary artery to the left lower lob.

MLAP: Mean left atrial pressure in cm H_2O, measured from the left atrium.

EMPAP: Effective mean pulmonary arterial pressure in cm H_2O, mean pulmonary arterial pressure minus mean left atrial pressure.

LVP: Left ventricular pressure in mm Hg, measured from a catheter in the left ventricular cavity inserted through the cardiac apex.

LVEDP: Left ventricular end-diastolic pressure in mm Hg, measured from the left ventricular pressure.

dp/dt: Maximal rate of rise of left ventricular pressure in mm Hg/sec derived from the left ventricular pressure with a derivative computer.

MAP: Mean aortic pressure in mm Hg, measured from a catheter inserted through a carotid artery.

MPAF: Mean pulmonary arterial flow in ml/min, measured with a Statham electromagnetic flow probe around the main pulmonary artery, using a Statham electromagnetic flowmeter.

HR: Heart rate, in beats/min, computed from aortic pressure waves, taken at a paper speed of 50 mm/sec.

Systemic vascular resistance: In Dynes·sec/cm^5; the quotient of mean aortic pressure minus left ventricular end-diastolic pressure in dynes/cm^2 and of mean pulmonary arterial flow in ml/sec.

Pulmonary vascular resistance: In dynes·sec/cm^5; the quotient of mean pulmonary arterial pressure minus mean left atrial pressure in dynes/cm^2 and of mean pulmonary arterial flow in ml/sec.

SV: Stroke volume: in ml, computed from cardiac output and heart rate.

SW: Stroke work: in gram·meter (g·m), computed from stroke volume and aortic pressure.

C. Results

The first group of experiments on spontaneously respiring dogs were designed to characterize the effects of administering TCE on the upper respiratory tract only, and on the lower respiratory tract bypassing the upper. The two routes of administration were alternately used in each dog, with 10 min of room air prior to each concentration of TCE. The results summarized in Tables 9.1 and 9.2 show a lack of consistency in control levels of the parameters of respiration and circulation. The variability in control measurements can be explained by the unpredictability of the respiratory effects of TCE, either a stimulation or a depression or both. As a result of the respiratory response, there was a variable uptake of TCE in the blood during the administration of TCE and also an unpredictable rate of elimination of TCE while the dog was breathing room air. Although the blood levels of TCE were not measured, experience with human subjects undergoing general anesthesia indicate variability in blood levels in situations similar to the conditions of our experiments.

TABLE 9.1

Bronchopulmonary Effects of Trichloroethylene (TCE) on the Upper Respiratory Tract of Anesthetized Closed-chest Dogs*

Procedure	Concentration	Respiratory rate (per min)			Respiratory minute volume (ml)			Mean aortic pressure (mm Hg)			Heart rate (per min)			Pulmonary compliance (ml/cm H_2O)			Pulmonary resistance (cm H_2O/LPS)		
	(% v/v)	C	E	Δ%	C	E	Δ%	C	E	Δ%	C	E	Δ%	C	E	Δ%	C	E	Δ%
Exposure of the upper resp. tract to trichloroethylene	0.1	18.3	17.0	−9.2	2314	2039	−12.2	150.8	148.3	−1.6	105	106.3	+1.1	41.5	48.6	+16.6	3.1	2.6	−15.9
		±2.66	±3.34	±6.33	±371.4	±422.2	±9.95	±5.22	±3.84	±1.00	±14.2	±14.55	±1.69	±4.38	±6.57	±8.36	±0.18	±0.20	±5.45
			NS			NS			NS			NS			0.10			0.02	
	0.5	16.8	16.8	+0.3	1951	1868	−5.6	151.3	153.5	−1.5	108.3	107.5	−0.7	51.0	51.5	−1.7	3.0	2.9	−2.9
		±3.17	±3.07	±1.95	±435.2	±450.6	±4.37	±3.15	±4.01	±0.96	±12.30	±12.22	±2.16	±3.92	±5.28	±9.33	±0.19	±0.28	±7.87
			NS			NS			NS			NS			NS			NS	
	1.0	15.5	15.3	−0.6	1810	1645	−7.8	152.8	150.8	−1.3	108.5	107.8	−2.7	48.2	51.4	+7.9	3.1	2.9	−5.6
		±2.90	±2.43	±4.14	±459.2	±400.88	±3.79	±2.78	±4.05	±1.40	±14.91	±13.16	±12.77	±5.47	±4.38	±6.21	±0.23	±0.14	±5.32
			NS			NS			NS			NS			NS			NS	
	5.0	16.8	13.5	−18.4	2184	1920	−14.8	150.8	152.3	+0.98	103.8	120.5	+18.8	54.2	53.2	−3.2	3.0	3.2	+4.4
		±2.53	±1.85	±4.25	±328.4	±412.1	±9.76	±3.38	±3.61	±4.05	±12.48	±8.63	±10.02	±6.07	±8.71	±6.82	±0.18	±0.34	±7.79
			0.02			NS			NS			0.10			NS			NS	

*Each group of data represents mean ±SE of the control (C) and experimental (E) values for 6 dogs, mean ±SE of Δ%, and significance level as p value (NS = not significant).

TABLE 9.2

Bronchopulmonary Effects of Trichloroethylene (TCE) on the Lower Respiratory Tract of Anesthetized Closed-chest Dogs*

Procedure	Concentration	Respiratory rate (per min)			Respiratory minute volume (ml)			Mean aortic pressure (mm Hg)			Heart rate (per min)			Pulmonary compliance (ml/cm H_2O)			Pulmonary resistance (cm H_2O/LPS)		
	(% v/v)	C	E	Δ%	C	E	Δ%	C	E	Δ%	C	E	Δ%	C	E	Δ%	C	E	Δ%
Trichloroethylene inhalation	0.1	16.5	17.8	+10.4	2540	2399	−6.9	146	139	−5.1	114	117	+3.7	48.3	44.3	−10.2	3.1	3.5	+12.7
		±2.18	±1.89	±10.09	±544.7	±614.0	±6.73	±10.9	±8.6	±4.34	±10.9	±8.6	±4.34	±6.65	±8.76	±5.73	±0.23	±0.25	±6.25
			NS			NS			0.05			NS			NS			0.1	
	0.5	17.0	16.8	−1.9	2605	2610	−1.9	139	146	+5.5	126	124	−1.5	45.1	44.6	−2.6	3.5	3.7	+6.4
		±2.35	±2.84	±5.99	±679.7	±751.6	±3.46	±8.75	±6.10	±2.84	±7.13	±9.11	±1.91	±5.95	±8.89	±9.22	±0.49	±0.46	±11.7
			NS			NS			0.05			NS			NS			0.1	
	1.0	17.8	17.0	−3.1	2393	2375	−9.5	145	144	−0.2	115	126	+10.7	40.2	44.0	+8.8	3.2	3.0	−7.3
		±1.93	±1.73	±7.52	±623.7	±816.6	±12.4	±4.50	±4.73	±0.83	±11.88	±9.24	±5.4	±3.47	±5.00	±4.55	±0.13	±0.23	±4.70
			NS			NS			NS			0.1			0.05			NS	
	5.0	17.0	18.3	+9.5	2773	2850	+0.6	140	137	−1.7	116	141	+24.3	45.2	43.4	−5.0	3.3	3.5	+6.7
		±2.00	±2.25	±14.99	±643.4	±786.3	±6.54	±4.50	±6.56	±2.96	±12.8	±6.57	±8.09	±4.40	±5.97	±5.05	±0.13	±0.11	±5.64
			NS			NS			NS			0.02			NS			NS	
Trichloroethylene after vagotomy	0.1	10.8	10.0	−6.8	2181	2295	+5.7	149	146	−2.0	198	200	+1.0	45.3	50.7	+19.4	3.3	2.9	−12.5
		±1.55	±1.68	±6.83	±1105.7	±195.3	±9.00	±7.49	±5.22	±1.39	±18.05	±18.0	±0.42	±7.09	±3.06	±16.28	±0.11	±0.33	±9.13
			NS			NS			NS			0.05			NS			NS	
	0.5	8.5	10.3	+27.4	1858	2443	+24.2	147	146	−0.8	197	202	−2.0	45.0	41.8	−6.4	3.1	3.5	+9.2
		±1.71	±1.60	±17.87	±184.5	±243.1	±23.9	±6.38	±6.10	±1.92	±19.28	±20.83	±1.24	±5.46	±5.93	±9.39	±0.40	±0.78	±12.5
			NS			NS			NS			0.1			NS			NS	
	1.0	10.3	12.0	+22.1	2405	2745	+11.0	147	145	−1.8	195	199	+1.6	48.8	49.5	+7.3	2.8	3.1	+8.6
		±1.80	±2.27	±22.2	±354.13	±106.04	±11.50	±6.38	±7.92	±1.66	±18.46	±20.30	±1.42	±4.73	±7.66	±25.5	±0.25	±0.74	±22.57
			NS			NS			NS			NS			NS			NS	
	5.0	9.0	12.3	+30.9	2035	1950	+4.9	149	140	−6.2	189	177	−6.6	54.0	51.4	−2.4	2.9	3.1	+6.3
		±1.08	±3.35	±30.13	±211.68	±436.6	±21.9	±6.42	±6.93	±2.26	±17.61	±20.77	±3.92	±5.73	±3.85	±11.06	±0.32	±0.46	±11.46
			NS			NS			0.05			0.01			NS			NS	

*Each group of data represents mean ±SE of the control (C) and experimental (E) values for 6 dogs, mean ±SE of Δ%, and significance level as p value (NS = not significant).

Even with a lack of uniformity in control values from dog to dog, and in the same dog from inhalation to inhalation, it was still possible to detect some consistency in responses to TCE; these are discussed in the next paragraphs.

Exposure of upper respiratory tract – The responses to increasing concentrations of TCE in the upper respiratory tract are summarized in Table 9.1. The most conspicuous effects were reduction in respiratory rate and production of tachycardia seen when 5.0% TCE was flushed in the upper respiratory tract. The lowest concentration used was 0.1% which caused a decrease in pulmonary resistance and an increase in compliance. However, higher concentrations did not influence pulmonary mechanics. The results suggest that the receptors influencing the caliber of the airways have a lower threshold for depression than those that mediate bradypnea.

Exposure of lower respiratory tract – The inhalation of 0.1% TCE caused no significant effect on respiratory rate, minute volume, pulmonary resistance, and compliance. Mean aortic blood pressure was either reduced at 0.1%, increased at 0.5% or unchanged at 1.0 and 5%. Heart rate was accelerated only at 1.0 or 5.0% (Table 9.2).

Repetition of the inhalation of TCE after cervical vagotomy revealed the role of the vagus in mediating the initial responses. They are as follows: (a) The fall in aortic blood pressure seen with the lowest concentration (0.1%) disappeared after vagotomy which also caused the appearance of hypotension from inhalation of the highest concentration (5.0%) of TCE. (b) The tachycardia seen with vagi intact disappeared after vagotomy.

Hemodynamic effects in dogs under artificial respiration – The second group of dogs were breathing artificially by a pump that forced either air or mixtures of TCE. The control levels summarized in Table 9.3 are less variable than those in the spontaneously breathing dog. It was, therefore, possible in the experiments with controlled respiration to obtain dose-related responses. The first noticeable effect was observed on the contractility of the heart as reflected by the decrease in the maximal rate of rise of left ventricular pressure (dp/dt). Progressively increasing concentrations of TCE administered by inhalation brought about a progressive decrease in peak left ventricular pressure (LVP), maximal rate of rise of left ventricular pressure (dp/dt), aortic pressure (AP), and mean pulmonary arterial flow (MPAF). A significant decrease in stroke volume (SV) and in stroke work (SW) was observed with 0.5 and 1.0% concentrations of TCE. Table 9.3 summarizes the various changes in the hemodynamics following inhalation of various concentrations of TCE. A record of a typical experiment is depicted in Figure 9.1.

Inhalation of TCE in a concentration of 0.01% for a period of 5 min brought about no significant change in any of the measured parameters. The only significant change observed after inhalation of 0.05% TCE was a decrease in the left ventricular dp/dt averaging 7.4% as compared with control mean level. Inhalation of 0.10% TCE resulted in a decrease in peak left ventricular pressure, dp/dt, and mean pulmonary arterial flow by 5.7%, 8.7%, and 3.8%, respectively. In a concentration of 0.25%, TCE brought about the following percent changes compared with the average control level: 6.4% decrease in peak left ventricular pressure, 13.5% decrease in left ventricular dp/dt, 6.2% decrease in mean aortic pressure, and 7.0% decrease in mean pulmonary arterial flow. A significant reduction in the various parameters was observed after inhalation of 0.5% TCE, namely, 11.6% in left ventricular pressure, 17.5% in left ventricular dp/dt, 8.4% in mean arterial pressure, 9.3% in mean pulmonary arterial flow, 9.7% in stroke volume, and 17.6% in stroke work. Following the inhalation of 1.0% TCE, there was a decrease in peak left ventricular pressure, left ventricular dp/dt, mean arterial pressure, mean pulmonary arterial flow, stroke volume, and stroke work of 17.3%, 25.5%, 17.1%, 15.1%, 14.9%, and 25.5%, respectively. No significant change was noted, however, with mean pulmonary arterial pressure, mean left atrial pressure, effective mean pulmonary arterial pressure, left ventricular end-diastolic pressure, heart rate, and both systemic and pulmonary vascular resistance.

Interaction between TCE and FC 11 – Trichlorofluoromethane (FC 11) in a concentration of 0.5% was found to produce the following mean percent decrease in the respective parameters (Table 9.4): 6.4% in peak left ventricular pressure, 8.6% in dp/dt, 6.2% in mean arterial pressure, 6.6% in mean pulmonary arterial flow, 7.9% in stroke volume, and 12.2% in stroke work. Inhalation of a mixture of 0.5% trichlorofluoromethane (FC 11) and 0.05% TCE led to the following decrease in the same parameters: 7.7%, 10.8%, 5.3%, 7.5%, 9.5%, and 14.5%, respectively. These

TABLE 9.3

Hemodynamic Responses to Inhalation of Trichloroethylene (TCE) in Anesthetized Open-chest Dogs*

Procedure	MPAP cm H_2O		MLAP cm H_2O		EMPAP cm H_2O		LVP mm Hg		LVEDP mm Hg		dp/dt mm Hg/Sec		MAP mm Hg		MPAF ml/min		HR beats/min		Vascular resistance dynes. sec/cm^5 Systemic		Vascular resistance dynes. sec/cm^5 Pulmonary		Stroke vol. ml		Stroke work g·m	
	C	E	C	E	C	E	C	E	C	E	C	E	C	E	C	E	C	E	C	E	C	E	C	E	C	E
0.01% TCE	46.5	44.0	11.5	11.5	35.0	32.5	119	120	3	5	4625	4750	110	109	1950	1950	148	145	4004	3696	1412	1306	10	10	14.7	14.3
	-2.5 ± 2.5		0.0 ± 0.0		-2.5 ± 3.5		+1.0 ± 0.5		+2.0 ± 3.5		+125 ± 125		-1.0 ± 1.0		0.0 ± 0.0		-3.0 ± 4.5		-308 ± 148		-106 ± 146		0.0 ± 0.0		-0.2 ± 0.13	
	NS		NS		NS		NS		NS		NS		NS		NS		NS		NS		NS		NS		NS	
0.05% TCE	37.4	34.6	10.2	10.2	26.8	23.8	120	118	5	7	4292	4000	109	106	2125	2092	165	166	4317	4097	915	851	14.2	13.8	22.8	22.2
	-2.8 ± 1.5		0.0 ± 0.63		-3.0 ± 1.8		-2.0 ± 2.14		2.0 ± 3.7		-292 ± 76.8		-3.0 ± 2.2		-33 ± 16.7		+1.0 ± 1.3		-220 ± 154.5		-64 ±68.8		-0.4 ± 0.25		-0.6 ± 0.51	
	NS		NS		NS		NS		NS		0.01		NS		NS		NS		NS		NS		NS		NS	
0.10% TCE	40.2	40.6	10.0	10.2	30.0	30.2	121	117	7	9	3917	3594	110	104	2075	2008	166	168	4540	4211	1024	1116	14.0	13.2	22.9	21.0
	+0.04 ± 0.9		+0.2 ± 0.9		+0.2 ± 0.7		-7.0 ± 3.6		+2.0 ± 1.67		-323.0 ± 91.9		-6.0 ± 2.9		-67.0 ± 30.9		+2.0 ± 2.3		-330 ± 304.2		+92 ± 49.5		-0.8 ± 0.4		-1.9 ± 0.79	
	NS		NS		NS		NS		NS		0.01		NS		0.02		NS		NS		NS		NS		NS	
0.25% TCE	37.8	36.0	9.5	8.8	28.2	27.2	121	113	5	7	4250	3708	111	103	2083	1958	160	160	4644	4572	1039	1129	14.4	13.6	23.0	20.9
	-1.8 ± 1.4		-0.7 ± 0.9		-1.0 ± 2.6		-8.0 ± 2.6		-2.0 ± 0.8		-542 ± 105.4		-8.0 ± 4.0		-125.0 ± 21.9		0.0 ± 4.0		-72.0 ± 107.2		+89 ± 112.3		-0.8 ± 0.7		-2.1 ± 0.90	
	NS		NS		NS		0.01		NS		0.001		0.01		0.001		NS		NS		NS		NS		0.02	
0.05% TCE	38.0	31.4	10.8	11.5	26.8	21.4	122	108	6	7	4146	3417	109	100	1983	1850	160	163	4875	4747	1002	806	13.8	12.6	22.7	19.4
	-6.6 ± 1.9		+0.7 ± 0.5		-5.4 ± 2.3		-14 ± 4.3		+1.0 ± 1.4		-724 ± 181.8		-9.0 ± 4.7		-133 ± 44.4		+3.0 ± 2.7		-128 ± 179.6		-197 ± 145.9		-1.2 ± 0.4		-3.3 ± 0.80	
	NS		NS		NS		0.01		NS		0.001		0.01		0.001		NS		NS		NS		0.01		0.01	
1.0% TCE	39.2	34.8	7.1	7.3	32.6	27.9	120	99	4	7	4667	3479	110	91	2100	1825	158	160	4347	3775	1146	1280	14.6	12.6	23.9	18.1
	-4.4 ± 3.9		-0.2 ± 0.9		-4.7 ± 3.5		-21 ± 5.9		+3 ± 1.5		-1188 ± 304.1		-19 ± 7.9		-275 ±38.4		+2 ± 3.3		-571 ± 379		-134 ± 196		-2 ± 0.3		-5.7 ± 0.9	
	NS		NS		NS		0.01		NS		0.001		0.01		0.001		NS		NS		NS		0 001		0.001	

*Each group of data represents mean of the control (C) and experimental (E) values for 10 dogs, mean ±SE of the difference, and significance level as P value (NS = not significant).

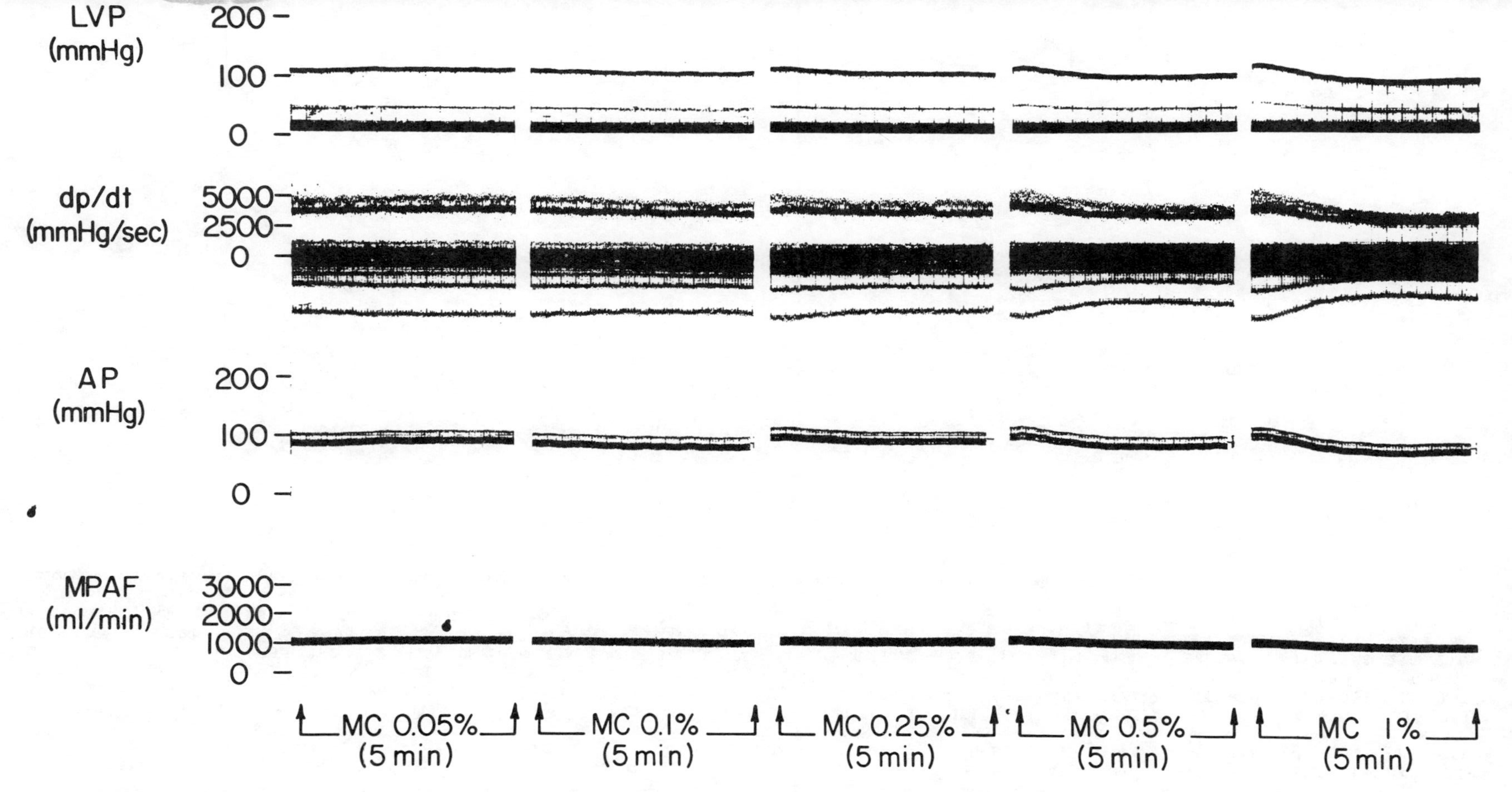

FIGURE 9.1 Recordings of left ventricular pressure (LVP), rate of rise of pressure (dp/dt), aortic blood pressure (AP), and mean pulmonary arterial flow (MPAF).

TABLE 9.4

Hemodynamic Responses to Inhalation of Trichloroethylene (TCE) and Trichlorofluoromethane (FC 11) Anesthetized Open-chest Dogs*

Procedure	MPAP cm H_2O		MLAP cm H_2O		EMPAP cm H_2O		LVP mm Hg		LVEDP mm Hg		dp/dt mm Hg/sec		MAP mm Hg		MPAF ml/min		HR beats/min		Vascular resistance dynes. sec/cm^5 Systemic		Vascular resistance dynes. sec/cm^5 Pulmonary		Stroke vol. ml		Stroke work g·m	
	C	E	C	E	C	E	C	E	C	E	C	E	C	E	C	E	C	E	C	E	C	E	C	E	C	E
0.05% TCE	37.4	34.6	10.2	10.2	26.8	23.8	120	118	5	7	4292	4000	109	106	2125	2092	165	166	4317	4097	915	851	14.2	13.8	22.8	22.2
	−2.8 ± 1.50		0.0 ± 0.63		−3.0 ± 1.8		−2.0 ± 2.14		2.0 ± 3.7		−292 ± 76.8		−3.0 ± 22		−33 ± 16.7		+1.0 ± 1.3		−220 ± 154.5		−64 ± 68.8		−0.4 ± 0.25		−0.6 ± 0.51	
	NS		NS		NS		NS		NS		0.01		NS		NS		NS		NS		NS		NS		NS	
FC 11/0.5%	35.4	33.8	11.3	12.2	25.0	22.8	119	112	5	5	4636	4198	111	105	2017	1900	154	154	45545	45057	918	925	14.6	13.6	23.6	21.1
	+1.6 ± 0.75		+0.8 ± 0.40		−22 ± 0.86		−7.3 ± 2.17		+0.2 ± 0.4		−437.5 ± 182.5		−6.3 ± 2.00		−116.7 ± 26.50		−0.2 ± 1.51		−48.8 ± 77.54		+7.4 ± 38.0		−1.0 ± 0.32		−2.5 ± 0.63	
	NS		NS		NS		0.01		NS		NS		0.01		0.01		NS		NS		NS		0.02		0.01	
Mix	39.6	35.6	13.2	13.5	27.2	23.2	116	108	6	8	4271	3782	108	104	2000	1879	152	152	4371	4354	981	951	15.0	13.8	23.7	20.8
0.05% TCE	−4.0 ± 2.22		+0.3 ± 0.76		−4.0 ± 1.3		−8.3 ± 3.02		+1.7 ± 0.99		−489.5 ± 221.0		−4.3 ± 1.8		−120.8 ± 36.18		+0.2 ± 1.98		−17.3 ± 41.40		−29.8 ± 69.39		−1.2 ± 0.2		−2.9 ± 0.84	
0.5% FC 11	NS		NS		NS		0.02		NS		0.02		0.01		0.01		NS		NS		NS		0.01		0.01	

*Each group of data represents mean of the control (C) and experimental (E) values for 10 dogs, mean ±SE of the difference, and significance as p value (NS = not significant).

results indicate a lack of potential in the cardiac effects of the two inhalants.

D. Discussion

The results of the tests on spontaneously breathing dogs are similar to those observed in unanesthetized mice (Chapter 8). The initial exposure to TCE influenced breathing which resulted in an unpredictable rate of uptake of TCE in the blood so that it was not possible to obtain definitive conclusions on the electrocardiographic changes in mice and the bronchopulmonary effects in dogs. The experiments summarized in Table 9.1 indicate some trends on the biphasic action of TCE on the upper respiratory tract. A low concentration (0.1%) causes a decrease in airway resistance which could not be elicited with higher concentrations, probably because of blockade of the receptors in the upper respiratory tract. On the other hand, a high concentration of TCE (5.0%) elicits respiratory depression that is resistant to blockade. These results indicate the presence of at least two types of receptors in the upper respiratory tract, one influencing the airways and the other the respiratory center.

The administration of low concentrations of TCE in the lower respiratory tract leads to activation of vagal afferents which causes a fall in aortic blood pressure and tachycardia. With vagi cut, the effects of high concentrations of TCE are still manifested as a fall in aortic blood pressure. It should be noted that the hypotension which persists after vagotomy is caused by the absorption of TCE reaching the heart and blood vessels, the details of which have been identified in the dogs under artificial respiration.

The present investigation demonstrates that TCE has no effect on the anesthetized open-chest dog preparation in 0.01% concentration. The myocardial depressant effect of TCE, as indicated from the decrease in the maximal rate of rise of the left ventricular pressure, was observed with doses as low as 0.05%. It is evident that the most sensitive index of TCE effect is the decrease in myocardial contractility. This effect is dose-dependent, increasing in magnitude when a higher concentration of TCE is administered. Thus, 0.05%, 0.10%, 0.25%, 0.5%, and 1.0% concentrations of TCE brought about a decrease in myocardial contractility of 7.4%, 8.7%, 13.5%, 17.5%, and 25.5%, respectively. There was a decrease in cardiac output, as indicated from the decrease in pulmonary arterial blood flow, amounting to 1.9%, 3.8%, 7.0%, 9.3%, and 15.1% with the various concentrations of TCE mentioned above. Meanwhile, there was a concomitant decrease in mean arterial pressure averaging 2.7%, 5.3%, 6.2%, 8.4%, and 17.1% after inhalation of increasing concentrations of TCE. No significant change was observed in either the systemic or pulmonary vascular resistances.

E. Summary

The administration of increasing concentrations of TCE to anesthetized dogs caused the following effects: (1) The minimal inhaled active concentration of 0.05% (2.67 mg/l or 500 ppm) caused a depression in myocardial contractility. (2) The depression of the heart is accompanied by a reduction in pulmonary blood flow with inhalation of 0.1% (5.35 mg/l or 1,000 ppm). (3) Higher concentrations ranging from 0.25 to 1.0% (13.14 to 53.5 mg/l or 2,500 to 10,000 ppm) caused tachycardia and systemic hypotension which is mediated, in part, through vagal afferents and, in part, as a manifestation of a reduction in cardiac output. (4) The highest concentration of 5.0% (267.5 mg/l or 50,000 ppm) caused severe depression of myocardial contractility and hypotension, all manifestations of direct effects of TCE on the heart and blood vessels. In the upper respiratory tract, TCE at 5.0% concentration stimulates receptors which depress respiratory rate and accelerate heart rate. (5) The changes in pulmonary mechanics were not significant at the various concentrations administered to the upper or lower respiratory tract.

Chapter 10

CONCLUDING REMARKS ON COMPARATIVE TOXICITY OF METHYL CHLOROFORM AND TRICHLOROETHYLENE

This monograph was prepared in response to the question of the possibility of health hazards resulting from exposure to trichloroethylene (TCE), specifically when contained in consumer products. As expected, there is information on acute intoxication caused by TCE, i.e., as an abused inhalant, or as an ingested poison, or as an anesthetic. However, we found no data in the literature on accidental consumer toxicity due to aerosols containing TCE. Additional investigation is needed to determine if any toxicity occurs in the course of the normal use of aerosol products. The ultimate answer will depend on the availability of several types of investigation: (a) epidemiologic survey of diseases in users of products containing TCE and in workers exposed to TCE; (b) concentration of TCE in the environment of the user; and (c) toxicologic investigation in animals of TCE alone and in combination with propellants and the constituents dispensed in the aerosol. The experiments reported in Chapters 8 and 9 belong to (c) and are examples of techniques for identifying the cardiovascular and bronchopulmonary effects of TCE in two animal species. When (c) is completed together with (a) and (b), additional data will still be needed to establish a cause-and-effect relationship between TCE and the disease. The literature on occupational exposure to TCE reviewed in Chapter 7 contains several reports of diseases occurring in workers exposed to levels probably above the Threshold Limit Value. The identification of the cause of the disease is complicated by the exposure of workers to other pollutants or by the fact that the worker himself may have a constitutional predisposition to the disease. The current problem of implicating TCE as the cause of an occupational disease will reappear in the future even though efforts are made to complete epidemiologic and experimental investigations.

These monographs on methyl chloroform and TCE appearing in the same volume have offered an opportunity to compare the toxicity of both solvents. Both are trichlorinated hydrocarbons, one saturated and the other unsaturated, and both have many similar uses in industry and in consumer products. The following list indicates that the toxicity of methyl chloroform is less than that of TCE.

A. Lethal Dosages in Animals

All the available comparisons indicate that TCE is more toxic than methyl chloroform. The lethal doses for methyl chloroform are larger than those for TCE and the comparative dose ratios are as follows: 2.4 to 1 orally, 1.8 to 1 intraperitoneally, and 1.5 to 1 inhalationally in rats; 4.9 to 1 orally, 1.9 to 1 intraperitoneally, and 1.5 to 1 inhalationally in mice (see Chapter 8). The differences in lethal dosage were first recognized by Lazarew in 1929 but little comparative work has been done since then.

B. Cardiotoxicity in Animals

The threshold concentrations that depressed the contractility of the canine heart are as follows: methyl chloroform 0.25% and TCE 0.05%, a ratio of 5 to 1 (see Chapter 9). On this basis, the cardiac effect of TCE as a solvent is four times greater than trichloromonofluoromethane (FC 11), the most toxic among aerosol fluorinated propellants. However, there was no potentiation of cardiac depression when both inhalants were administered simultaneously.

C. General Anesthesia

It is taken for granted that TCE is a general anesthetic whereas methyl chloroform is not. In reality, both are general anesthetics and the potency of methyl chloroform is lower by a ratio of 1.8 to 1. It appears, from inhalational experiments with TCE in mice, that death is caused by central nervous depression manifested by respiratory arrest. The margin of safety of TCE as an anesthetic is narrower than that of methyl chloroform (see Chapter 7).

D. Hepatotoxicity

In animals both solvents are potentially toxic to the liver but the dose of methyl chloroform is higher than that of TCE by a factor of 1.3 to 1 (see Chapter 7).

E. Threshold Limit Values

The limits set for the two solvents are as follows: methyl chloroform 350 ppm and TCE 100 ppm.

F. Incidence of Abuse

The literature on abused inhalants indicates that TCE is more frequently used than methyl chloroform. This is true not only in the United States but also in Europe and the Far East. It appears that TCE may have an effect known to teenagers that encourages them to abuse the solvent.

G. Incidence of Occupational Disease

The literature on this subject shows a predominance of poisoning caused by TCE. A rough approximation indicates that there are probably ten times more incidents of poisoning by TCE than by methyl chloroform (see Chapter 7). The reasons for the higher incidence could include greater use of TCE in industry as well as a higher toxicity of TCE. Additional investigation is needed to determine the causes for the higher incidence of poisoning by TCE and to develop measures to reduce its concentration in the environment.

REFERENCES – Trichloroethylene

1. U.S. Department of Health, Education, and Welfare, *Occupational Exposure to Trichloroethylene*, U.S. Government Printing Office, Washington, D.C., 1973, 1-102.
2. **Fischer, E.**, Ueber die Einwirkung von Wasserstoff auf Einfach-Chlorkohlenstoff, *Jena Z. Med. Naturwiss.*, 1, 123, 1864.
3. **Plessner, W.**, Über Trigeminuserkrankung infolge von Trichlorethylenvergiftung, *Neurol. Zentralbl.*, 34, 916, 1915.
4. **von Oettingen, W. F.**, The halogenated hydrocarbons: their toxicity and potential dangers, *J. Ind. Hyg. Toxicol.*, 19, 349, 1937.
5. **von Oettingen, W. F.**, *The Halogenated Aliphatic, Olefinic, Cyclic, Aromatic, and Aliphatic-aromatic Hydrocarbons Including the Halogenated Insecticides, Their Toxicity and Potential Dangers*, U.S. Department of Health, Education, and Welfare, 203, Publ. No. 414, 1955.
6. **Browning, E.**, *Toxicity of Industrial Organic Solvents*, Her Majesty's Stationery Office, London, 169, 1953.
7. **Oljenick, I.**, Trichloroethylene treatment of trigeminal neuralgia, *J. Am. Med. Assoc.*, 91, 1085, 1928.
8. **Striker, C., Goldblatt, S., Wann, I. S., and Jackson, D. E.**, Clinical experiences with use of trichloroethylene in production of over 300 analgesias and anesthesias, *Anesth. Analg.*, Cleveland, 14, 68, 1935.
9. **Atkinson, R. S.**, Trichloroethylene anaesthesia, *Anesthesiology*, 21, 67, 1960.
10. **Defalque, R. J.**, Pharmacology and toxicology of trichloroethylene, *Clin. Pharmacol. Ther.*. 2, 665, 1961.
11. **Smith, G. F.**, Trichloroethylene: A review, *Br. J. Ind. Med.*, 23, 249, 1966.
12. **Noweir, M., Pfitzer, E. A., and Hatch, T. F.**, Decomposition of chlorinated hydrocarbons: a review, *Am. Ind. Hyg. Assoc. J.*, 33, 454, 1973.
13. **Stowell, T. E. A.**, Correspondence: Anaesthesia for laryngofissure, *Br. Med. J.*, 1, 302, 1943.
14. **Lehmann, K. B.**, Experimentelle Studien über den Einfluss technisch und hygienish Wichtigen Gase und Dämpfe auf den Organismus, *Arch. Hyg.*, 74, 1, 1911.
15. **McClelland, M.**, Some toxic effects following Trilene decomposition products, *Proc. R. Soc. Med.*, 37, 526, 1944.
16. **Jones, G. W. and Scott, G. S.**, U.S. Bureau of Mines – Report of Investigation 3666, Washington, D.C., 1942.
17. **Strayer, W. M.**, Some methods of analysis and determination of anesthetic agents, *Anesthesiology*, 11, 599, 1950.
18. **Stack, V. T., Jr., Forrest, D. E., and Wahl, K. K.**, Determination of trichloroethylene in air, *Am. Ind. Hyg. Assoc. J.*, 22, 184, 1961.
19. **Yuhi, K.**, Studies on the chemical estimation method of industrial environment. II. Determination of trichloroethylene, chloroform and benzylchloride by pyridine-formalin method, *Jpn. J. Hyg.*, 21, 407, 1967, (in Japanese).
20. **Hall, K. D., Garlington, L. N., Nowill, W. I., and Stephen, C. R.**, The analysis of small concentrations of trichloroethylene vapor by interferometry, *Anesthesiology*, 14, 38, 1953.
21. **Stewart, R. D. and Erley, D. S.**, Detection of volatile organic compounds and toxic gases in humans by rapid infrared techniques, in *Progress in Chemical Toxicology*, Stolman, A., Ed., Academic Press, New York, 2, 183, 1965.
22. **Sassé, R. A.**, Gamma and neutron radiolysis of the system trichloroethylene-oxygen-water, *Health Phys.*, 13, 1015, 1967.
23. **Simmons, J. H. and Moss, I. M.**, Measurement of personal exposure to 1,1,1-trichloroethane and trichloroethylene using an inexpensive sampling device and battery-operated pump, *Ann. Occup. Hyg.*, 16, 47, 1973.
24. **Morgan, D. J. and Duxbury, G.**, The determination of chlorinated hydrocarbons in the atmosphere by activation analysis, *Ann. Occup. Hyg.*, 8, 253, 1965.
25. **Fraust, C. L. and Hermann, E. R.**, Charcoal sampling tubes for organic vapor analysis by gas chromatography, *Am. Ind. Hyg. Assoc. J.*, 27, 68, 1966.
26. **Dobkin, A. B. and Byles, P. H.**, Trichloroethylene anesthesia, *Clin. Anesth.*, 1, 43, 1963.
27. **Parkhouse, J.**, Trichloroethylene, *Br. J. Anaesth.*, 37, 681, 1965.
28. **Chenoweth, M. B., Ed.**, *Modern Inhalation Anesthetics*, Springer-Verlag, Berlin, 1972, 116.
29. **Pembleton, W. E.**, Trichloroethylene anesthesia re-evaluated, *Anesth. Analg.* (Cleveland), 53, 730, 1974.
30. **Svandzhian, E. P.**, Trichloroethylene (narcogen) for anesthesia and analgesia, *Eksp. Khir. Anesteziol.*, 8, 60, 1963 (in Russian).
31. **Carmel, A. G.**, Trichloroethylene inhalant analgesia in office practice: experience in 6,000 administrations, *Dis. Colon Rectum*, 13, 138, 1970.
32. **Hovell, B. C., Masson, A. H. B., and Wilson, J.**, Trichloroethylene for post-operative analgesia. A clinical assessment of trichloroethylene for the relief of post-operative pain, *Anaesthesia*, 22, 284, 1967.
33. **Jončev, V. and Marinov, L.**, Narcohypnosis with trilene, *Folia. Med.* (Plovdiv), 11, 281, 1969.
34. **Néel, J. L. and Dupont, P.**, Etude du test du papier au trichloréthylène chez les sujets coronariens (sur 128 cas), *Arch. Mal. Coeur. Vaiss.* (Suppl. 3, Rev. Atheroscler.), 3, 46, 1961.
35. **Thrower, J. C.**, Iatrogenic effect of drugs and anesthesia, *J. S. C. Med. Assoc.*, 59, 1, 1963.
36. **Hargarten, J. J., Hetrick, G. H., and Fleming, A. J.**, Industrial safety experience with trichloroethylene: its use as a vapor degreasing solvent 1948–1957, *Arch. Environ. Health*, 3, 461, 1961.
37. **Dale, R. M.**, The control of solvent emission from solvent vapour cleaning plants, *Ann. Occup. Hyg.*, 15, 85, 1972.

38. **Perret, J., Arondel, E., Guyotjeannin, C., and Pitton, M.,** A propos de l'utilisation abusive d'une graisse courante, *Arch. Mal. Prof. Med. Trav. Secur. Soc.,* 33, 129, 1972.
39. **Dahlberg, J. A. and Myrin, L. M.,** The formation of dichloroacetyl chloride and phosgene from trichloroethylene in the atmosphere of welding shops, *Ann. Occup. Hyg.,* 14, 269, 1971.
40. **Rinzema, L. C.,** Behavior of chlorohydrocarbon solvents in the welding environment, *Int. Arch. Arbeitsmed.,* 28, 151, 1971.
41. **Glass, W. I., Harris, E. A., and Whitlock, R. M. L.,** Phosgene poisoning: case report, *N. Z. Med. J.,* 74, 386, 1971.
42. **Cary, W. H., Jr. and Hepler, J. M.,** Health hazards in the dry cleaning industry, *Am. J. Public Health,* 28, 1029, 1938.
43. **Bourne, R. G.,** Solvent hazards in dry-cleaning industry, *Med. J. Aust.,* 1, 879, 1955.
44. **Preite, E.,** Indagini sulle condizioni igienico-sanitarie di alcune lavanderie a secco della citta di Modena, *Riv. Ital. Ig.,* 25, 205, 1965.
45. **Brancaccio, A., Fermariello, U., and Castellino, N.,** Rischio lavorativo nelle lavanderie a secco con impiego di tricloroetilene, *Folia Med.* (Naples), 49, 729, 1966.
46. **Veljanovski, A. and Petrov, T.,** Prilog zaštiti radnika u hemijskim čistionicama, *Vojnosanit. Pregl.,* 29, 183, 1972.
47. **Truhaut, M. R., Boudène, C., Phy-Lich, N., and Catella, H.,** Contribution à l'étude du dosage du trichloréthylène dans les solvants complexes, *Arch. Mal. Prof. Med. Trav. Secur. Soc.,* 28, 369, 1967.
48. **Gleason, M. N., Gosselin, R. E., Hodge, H. C., and Smith, R. P.,** *Clinical Toxicology of Commercial Products. Acute Poisoning,* 3rd ed., Williams and Wilkins, Baltimore, 1969.
49. **Hoschek, R.,** Plötzliche Spättodesfälle nach geringfügiger Trichloräthylen-Einwirkung, *Int. Arch. Gewerbepathol. Gewerbehyg.,* 19, 319, 1962.
50. **Barni, M., Fabroni, F., and Gentili, M.,** La mort subite dans l'intoxication chronique par le trichlorethylene, *Med. Lav.,* 59, 425, 1968.
51. **Hristić, L. and Graovac-Leposavic, L.,** Zadesno trovanje trihloretilenom, *Med. Glas.,* 24, 376, 1970.
52. **Hübner, A.,** Aus Unfallakten. Einatmung von Tri-Dämpfen keine Berufskranheit, *Monatsschr. Unfallheikd. Versicher. Versorg. Verkehrsmed.,* 63, 26, 1960.
53. **Truhaut, R., Boudène, C., and Bister-Miel, F.,** Application d'un appareil portatif à la detérmination en continu de l'exposition individuelle au trichloréthylène dans l'industrie, *Arch. Mal. Prof. Med. Trav. Secur. Soc.,* 25, 577, 1964.
54. **Schollmeyer, W.,** Plötzlicher Tod durch Trichloräthylen-Vergiftung bei Einwirkung dieses Giftes über Längere Zeit, *Arch. Toxikol.,* 18, 229, 1960.
55. **Trense, E. and Zimmermann, H.,** Tödliche Inhalationsvergiftung durch chronisch einwirkende Perchloräthylen-dämpfe, *Zentralbl. Arbeitsmed. Arbeitsschutz.,* 19, 131, 1969.
56. **Huff, J. E.,** New evidence on the old problems of trichloroethylene, *Ind. Med.,* 40, 25, 1971.
57. **Antonyuzhenko, V. A., Kashtanova, I. M., and Khii, R. G.,** Features peculiar to clinical manifestations of a chronic poisoning with low concentrations of chlorinated carbohydrogestetrachloroethane and trichloroethylene, *Gig. Tr. Prof. Zabol.,* 9, 37, 1965 (in Russian).
58. **Hadengue, A., Facquet, J., Colvez, P., and Jullien, J. L.,** Intoxication par le nitrobenzène et le trichloréthylène, *Arch. Mal. Prof. Med. Trav. Secur. Soc.,* 21, 43, 1960.
59. **Bourret, J., Viallier, J., Tolot, F., and Robillard, J.,** Polynévrites par exposition simultanée au trichloréthylène et à l'essence, *Rev. Med. Suisse Romande,* 88, 173, 1968.
60. **Wende, E.,** Arbeitsmedizinische Aspekte der Gummi-Industrie, *Fortschr. Med.,* 90, 239, 1972.
61. **Zieliński, A.,** Narażenie zawodowe na olów i trojchloroetylen a palenie tytoniu na stanowisku pracy, *Przegl. Lek.,* 25, 656, 1969.
62. **Kanetaka, T. and Oka, T.,** Toxic liver injuries, *Acta Pathol. Jpn.,* 23, 617, 1973.
63. **Neubaur, E.,** Arbeitsmedizinische Aufgaben eines Fliegerarztes, *Fortschr. Med.,* 91, 129, 1973.
64. **Gutman, G.,** Accidental intoxication with trichlorethylene, *Harefuah,* 57, 238, 1959 (in Hebrew).
65. **Stewart, R. D., Hake, C. L., and Peterson, J. E.,** "Degreaser's flush"; dermal response to trichloroethylene and ethanol, *Arch. Environ. Health,* 29, 1, 1974.
66. **Gwynne, E. I.,** Trichloroethylene neuropathy (letter), *Br. Med. J.,* 2, 315, 1969.
67. **Seage, A. J. and Burns, M. W.,** Pulmonary oedema following exposure to trichlorethylene, *Med. J. Aust.,* 2, 484, 1971.
68. **Harenko, A.,** Two peculiar instances of psychotic disturbance in trichloroethylene poisoning, *Acta Neurol. Scand.,* 43(Suppl. 31), 139, 1967.
69. **Beisland, H. O. and Wannag, S. A.,** Trikloretylensniffing. Akutte lever-og nyreskader, *Tidsskr. Nor. Laegeforen.,* 90, 285, 1970.
70. **Clearfield, H. R.,** Hepatorenal toxicity from sniffing spot-remover (trichloroethylene), *Am. J. Dig. Dis.,* 15, 851, 1970.
71. **Baerg, R. D. and Kimberg, D. V.,** Centrilobular hepatic necrosis and acute renal failure in "solvent sniffers," *Ann. Intern. Med.,* 73, 713, 1970.
72. **Graovac-Leposavić, L., Milosavljević, Z., and Ilić, V.,** Ispitivanje funkcija jetre radnika eksponiranih trihloretilenu, *Arh. Hig. Rada Toksikol.,* 15, 93, 1964.
73. **Edh, M., Selerud, A., and Sjöberg, C.,** Dödsfall i samband med missbruk av organiska lösningsmedel, *Lakartidningen,* 70, 3949, 1973.

74. **Scherzer, E.**, Über eine seltene Trichloräthylenschädigung, *Psychiatr. Neurol.* (Basel), 148, 110, 1964.
75. **Musclow, C. E. and Awen, C. F.**, Glue sniffing: report of a fatal case, *Can. Med. Assoc. J.*, 104, 315, 1971.
76. **James, W. R. L.**, Fatal addiction to trichloroethylene, *Br. J. Ind. Med.*, 20, 47, 1963.
77. **Harrer, G., Kisser, W., Pilz, P., Sorgo, G., and Wölkart, N.**, Ergebnisse und Kasuistik. Über 3 Fälle von Trichloräthylen-bzw. Tetrachlorkohlenstoff-"Schnüffeln" mit letalem Ausgang, *Nervenarzt*, 44, 645, 1973.
78. **Migdal, A., Obodecka, Z., Ozog, L., Zarczyński, E., and Kraczek, T.**, Przypadek narkomanii inhalowania "tri" zakończony zgonem, *Wiad. Lek.*, 24, 471, 1971.
79. **Cragg, J. and Castledine, S. A.**, A fatality associated with trichloroethylene inhalation, *Med. Sci. Law*, 10, 112, 1970.
80. **Litt, I. F. and Cohen, M. I.**, "Danger . . . vapor harmful": spot-remover sniffing, *N. Engl. J. Med.*, 281, 543, 1969.
81. **Crooke, S. T.**, Solvent inhalation, *Tex. Med.*, 68, 67, 1972.
82. **Alapin, B.**, Trichlorethylene addiction and its effects, *Br. J. Addict.*, 68, 331, 1973.
83. **Rihoux, F. and Lachapelle, J. M.**, Deux cas de toxicomanie au trichlorethylene present dans un topique a usage dermatologique, *Arch. Belg. Dermatol. Syphiligr.*, 28, 409, 1972.
84. **Selden, A-M., Selerud, A., and Sjöberg, C.**, Thinnerrapport II: sniffning och brander, *Lakartidningen*, 70, 4049, 1973.
85. **Glatzel, J.**, Über eine Beobactung von Trichloräthylen-Sucht, *Psychiatr. Neurol. Med. Psychol.*, 19, 366, 1967.
86. **Luria, E. and Meneghini, R.**, Su di un caso di tossicomania da inalazione di tricloroetilene, *G. Psichiatr. Neuropatol.*, 93, 743, 1965.
87. **Guaraldi, G. P. and Bonasegla, F.**, Su di un caso di tossicomania da tricloroetilene, *Riv. Sper. Freniatr. Med. Leg. Alienazioni Ment.*, 92, 913, 1968.
88. **Hrynkiewicz, L.**, Przypadek narkomanii trojchloroetylenowej powiklanej objawami psychotycznymi, *Neurol. Neurochir. Psychiatr. Pol.*, 13, 941, 1963.
89. **Dmochowska-Mroczek, H.**, Ciezkie zaburzenia rytmu serca w przebiegu ostrego zatrucia trójchloroetylenem ("tri"), *Kardiol. Pol.*, 15, 343, 1972.
90. **Januszkiewicz-Grabias, A. and Szlabowicz, J.**, Problem narkomanii trójchloroetylenowej na podstawie obserwowanych przypadków, *Psychiatr. Pol.*, 4, 395, 1970.
91. **Škoricová, M. and Molčan, J.**, Catamnestic study on volatile solvent addiction, *Act. Nerv. Super.*, 14, 116, 1972.
92. **Škoricová, M.**, Katamnestická štúdia k výskytu osobitného druhu narkománii u mladistvých, *Cesk. Psychiatr.*, 68, 110, 1972.
93. **Dror, K.**, Poisoning due to aeroplane glue (letter), *Harefuah*, 84, 50, 1973 (in Hebrew).
94. **Pinkhas, J., Cohen, I., Kruglac, J., and De Vries, A.**, Poisoning caused by an airplane assembly hobby, *Harefuah*, 83, 326, 1973 (in Hebrew).
95. **Le Breton, R., Le Bourhis, J., and Garat, J.**, Un cas d'empoisonnement criminel par le trichloréthylène, *Ann. Med. Leg.*, 43, 281, 1963.
96. **Chiappino, G. and Secchi, G. C.**, Descrizione di un caso di intossicazione acuta da ingestione accidentale di 1,2-dicloropropano venduto come trielina, *Med. Lav.*, 59, 334, 1968.
97. **Secchi, G. C., Chiappino, G., Lotto, A., and Zurlo, N.**, Composizione chimica attuale delle trieline commerciali e loro effetti epatotossici. Studio clinico ed enzimologico, *Med. Lav.*, 59, 486, 1968.
98. **Curtarelli, G. and Grimaldi, M. G.**, Su un caso di epatite anitterica insorta dopo ingestione di tricloroetilene, *Med. Lav.*, 58, 301, 1967.
99. **Secchi, G. C. and Alessio, L.**, L'epatopatia acuta da ingestione di trieline del commercio: studio enzimologico, *Epatologia*, 17, 279, 1971.
100. **Gangl, A. and Pietschmann, H.**, Perorale Trichloräthylen-Intoxikation (Fallbericht), *Wien Klin. Wochenschr.*, 84, 675, 1972.
101. **Cohen-Solal, J.**, Intoxications par le trichloréthylène et les solvants, *Med. Infant.*, 70, 455, 1963.
102. **Berek-Pyzikowa, B. and Kamińska, D.**, Ostre zatrucia dzieci chemikaliami sluzacymi do czyszczenia, *Pediatr. Pol.*, 44, 459, 1969.
103. **Wiecko, W.**, Zatrucie trojchlorkiem etylenu (tri) zazytym doustnie, *Wiad. Lek.*, 19, 1117, 1966.
104. **Meyer, H-J.**, Perorale Vergiftung mit Trichloräthylen, *Arch. Toxikol.*, 21, 225, 1966.
105. **Migdal, A., Graczyk, E., Obodecka, Z., and Piesiak, K.**, Badania biochemiczne i neurologiczne w przebiegu doustnego zatrucia trójchloroetylenem ("tri"), *Wiad. Lek.*, 24, 1669, 1971.
106. **Borch, M.**, Spätfolgen einer Trichloräthylen-Tetrachlorkohlenstoffvergiftung im Kindesalter, *Psychiatr. Neurol. Med. Psychol.*, 25, 309, 1973.
107. **Nicholson, M. J.**, Accidental use of trichloroethylene (Trilene, Trimar) in a closed system. Case number 39, *Anesth. Analg.* (Cleveland), 43, 740, 1964.
108. **Lloyd, E. L.**, Trichloroethylene neuropathy (letter), *Br. Med. J.*, 2, 118, 1969.
109. **Radonov, D., Minceva, M., Mitev, L., and Lasarov, I.**, Accidents in women, treated with trichlorethylene, used for analgesia during induced interruption of pregnancy, *Akush. Ginekol.* (Sofiia), 12, 416, 1973 (in Russian).
110. **Capon, J. H.**, Atmospheric pollution with trichloroethylene in operating theatres (letter), *Anaesthesia*, 29, 96, 1974.
111. **Corbett, T. H., Hamilton, G. C., Yoon, M. K., and Endres, J. L.**, Occupational exposure of operating room personnel to trichloroethylene, *Can. Anaesth. Soc. J.*, 20, 675, 1973.

112. **Stopps, G. J. and McLaughlin, M.**, Psychophysiological testing of human subjects exposed to solvent vapors, *Am. Ind. Hyg. Assoc. J.*, 28, 43, 1967.
113. **Vernon, R. J. and Ferguson, R. K.**, Effects of trichloroethylene on visual-motor performance, *Arch. Environ. Health*, 18, 894, 1969.
114. **Ferguson, R. K. and Vernon, R. J.**, Trichloroethylene in combination with C. N. S. drugs, *Arch. Environ. Health*, 20, 462, 1970.
115. **Kylin, B., Axell, K., Samuel, H. E., and Lindborg, A.**, Effect of inhaled trichloroethylene on the C. N. S., *Arch. Environ. Health*, 15, 48, 1967.
116. **Stewart, R. D., Dodd, H. C., Gay, H. H., and Erley, D. S.**, Experimental human exposure to trichloroethylene, *Arch. Environ. Health*, 20, 64, 1970.
117. **Salvini, M., Binaschi, S., and Riva, M.**, Valutazione psicofisiologica nell'uomo del "threshold limit value" per il tricloroetilene, *Boll. Soc. Ital. Biol. Sper.*, 44, 1086, 1968.
118. **Salvini, M., Binaschi, S., and Riva, M.**, Evaluation of the psychophysiological functions in humans exposed to trichloroethylene, *Br. J. Ind. Med.*, 28, 293, 1971.
119. **Parkhouse, J.**, Uptake and metabolism of trichloroethylene, *Can. Anaesth. Soc. J.*, 16, 113, 1969.
120. **DuBois, A. B. and Rogers, R. M.**, Respiratory factors determining the tissue concentrations of inhaled toxic substances, *Respir. Physiol.*, 5, 34, 1968.
121. **Kimmerle, G. and Eben, A.**, Metabolism, excretion and toxicology of trichloroethylene after inhalation. 2. Experimental human exposure, *Arch. Toxikol.*, 30, 127, 1973.
122. **Stewart, R. D. and Dodd, H. C.**, Absorption of carbon tetrachloride, trichloroethylene, tetrachloroethylene, methylene chloride, and 1, 1, 1-trichloroethane through the human skin, *Am. Ind. Hyg. Assoc. J.*, 25, 439, 1964.
123. **Stewart, R. D., Gay, H. H., Erley, D. S., Hake, C. L., and Peterson, J. E.**, Observations on the concentrations of trichloroethylene in blood and expired air following exposure of humans, *Am. Ind. Hyg. Assoc. J.*, 23, 167, 1962.
124. **Stewart, R. D., Swank, J. D., Roberts, C. B., and Dodd, H. C.**, Detection of halogenated hydrocarbons in the expired air of human beings using the electron capture detector, *Nature*, 198, 696, 1963.
125. **Stewart, R. D., Hake, C. L., and Peterson, J. E.**, Use of breath analysis to monitor trichloroethylene exposures, *Arch. Environ. Health*, 29, 6, 1974.
126. **Prior, F. N.**, Blood levels of trichlorethylene during major surgery, *Anaesthesia*, 27, 379, 1972.
127. **Clayton, J. I. and Parkhouse, J.**, Blood trichloroethylene concentrations during anaesthesia under controlled conditions, *Br. J. Anaesth.*, 34, 141, 1962.
128. **Kelley, J. M. and Brown, B. R., Jr.**, Biotransformation of trichloroethylene, *Int. Anesthesiol. Clin.*, 12, 85, 1974.
129. **Sukhotina, K. L.**, Data on a study of metabolites of trichloroethylene in workers engaged in the production of trichloroethylene and monochloracetic acid, *Gig. Tr. Prof. Zabol.*, 13, 35, 1969 (in Russian).
130. **Sukhanova, V. A.**, Research data on trichloroethylene metabolism in adolescents trained for chemical plant machine operations, *Gig. Tr. Prof. Zabol.*, 15, 52, 1971 (in Russian).
131. **Nomiyama, K. and Nomiyama, H.**, Metabolism of trichloroethylene in humans. Sex difference in urinary excretion of trichloroacetic acid and trichloroethanol, *Int. Arch. Arbeitsmed.*, 28, 37, 1971.
132. **Bartoníček, V.**, The effect of some substances on the elimination of trichloroethylene metabolites, *Arch. Int. Pharmacodyn. Ther.*, 144, 69, 1963.
133. **Bartoníček, V. and Teisinger, J.**, Effect of tetraethyl thiuram disulphide (disulfiram) on metabolism of trichloroethylene in man, *Br. J. Ind. Med.*, 19, 216, 1962.
134. **Müller, G., Spassovski, M., and Henschler, D.**, Metabolism of trichloroethylene in man. II. Pharmacokinetics of metabolites, *Arch. Toxikol.*, 32, 283, 1974.
135. **Ertle, T., Henschler, D., Müller, G., and Spassowski, M.**, Metabolism of trichloroethylene in man. I. The significance of trichloroethanol in long-term exposure conditions, *Arch. Toxikol.*, 29, 171, 1972.
136. **Müller, G., Spassovski, M., and Henschler, D.**, Trichloroethylene exposure and trichloroethylene metabolites in urine and blood, *Arch. Toxikol.*, 29, 335, 1972.
137. **Bartoníček, V.**, Metabolism and excretion of trichloroethylene after inhalation by human subjects, *Br. J. Ind. Med.*, 19, 134, 1962.
138. **Ogata, M., Takatsuka, Y., and Tomokuni, K.**, Excretion of organic chlorine compounds in the urine of persons exposed to vapours of trichloroethylene and tetrachloroethylene, *Br. J. Ind. Med.*, 28, 386, 1971.
139. **Souček, B. and Vlachová, D.**, Excretion of trichloroethylene metabolites in human urine, *Br. J. Ind. Med.*, 17, 60, 1960.
140. **Lowry, L. K., Vandervort, R., and Polakoff, P. L.**, Biological indicators of occupational exposure to trichloroethylene, *J. Occup. Med.*, 16, 98, 1974.
141. **Elkins, H. B. and Pagnotto, L. D.**, Is the 24-hour urine sample a fallacy? *Am. Ind. Hyg. Assoc. J.*, 26, 456, 1965.
142. **Bardoděj, Z.**, Die Beurteilung der Trichloräthylenexposition mittels Harnanalyse, *Med. Welt*, 28, 1636, 1968.
143. **Herbolsheimer, R.**, Gaschromatographische Bestimmung von Trichloräthylen, Trichloräthanol, Trichloressigsäure und Äthanol in einem Analysengang aus einer Probe, *Arch. Toxikol.*, 32, 209, 1974.
144. **Lindner, J. and Langes, K.**, Detection of trichloroacetic acid and trichloroethanol in urine by headspace gas chromatography and conventional gas chromatography, *Mitteilungsbl. GDCh (Ges. Dtsch. Chem.)-Fachgruppe Lebensmittelchem. Gerichtl. Chem.*, 28, 163, 1974 (in German).

145. **Rivoire, J., Genevois, M., and Tolot, F.**, L'élimination de l'acide trichloracétique chez les sujets exposés au trichlorethylène. Résultats de l'emploi systématique de cette méthode dans une entreprise, *Arch. Mal. Prof. Med. Trav. Secur. Soc.*, 23, 395, 1962.
146. **Perret, J.**, La recherche de l'acide trichloracétique dans les urines des sujets exposés au trichloréthylène, *Arch. Mal. Prof. Med. Trav. Secur. Soc.*, 24, 676, 1963.
147. **Castaing, R., Liermain, A., Ferrus, L., Cardinaud, J-P., and Favarel-Garrigues, J-C.**, Intoxication aiguë par le trichloréthylène; A propos de 6 observations. Etude experimentale de l'elimination du produit, *J. Med. Bordeaux Sud-Ouest,* 144, 549, 1967.
148. **Rubino, G. F., Scansetti, G., Trompeo, G., and Gaido, P. C.**, Studio sull'intossicazione cronica da trielina. 4. II catabolismo dei derivati del tricloroetilene, *Med. Lav.*, 50, 755, 1959.
149. **Scansetti, G., Rubino, G. F., and Trompeo, G.**, Studio sull'intossicazione cronica da trielina. 3. Metabolismo del tricloroetilene, *Med. Lav.*, 50, 743, 1959.
150. **Rubino, G. F., Scansetti, G., and Trompeo, G.**, Studio sull'intossicazione cronica da trielina. 2. Assorbimento del tricloroetilene, *Med. Lav.*, 50, 733, 1959.
151. **Abrahamsen, A. M.**, Quantitative estimation of trichloroacetic acid in the urine and serum in trichloroethylene poisoning, *Acta. Pharmacol. Toxicol.*, 17, 288, 1960.
152. **Weichardt, H. and Bardodej, Z.**, Die Bestimmung von Trichloressigsäure in Urin von Tri-Arbeitern, *Zentralbl. Arbeitsmed. Arbeitsschutz.*, 20 (7), 219, 1970.
153. **Kulesza, K.**, Modified method for determination of trichloroethanol in urine, *Bull. Inst. Mar. Med. Gdansk.*, 15, 207, 1964.
154. **Beliakov, A. A.**, Determination of trichloroethylene and tetrachloroethane in the air and their metabolites (trichloroacetic acid and chloroform) in the urine, *Gig. Tr. Prof. Zabol.*, 10, 51, 1966 (in Russian).
155. **Ikeda, M., Ohtsuji, H., Kawai, H., and Kuniyoshi, M.**, Excretion kinetics of urinary metabolites in a patient addicted to trichloroethylene, *Br. J. Ind. Med.*, 28, 203, 1971.
156. **Nomiyama, K.**, Estimation of trichloroethylene exposure by biological materials, *Int. Arch. Arbeitsmed.*, 27, 281, 1971.
157. **Imamura, T. and Ikeda, M.**, Lower fiducial limit of urinary metabolite level as an index of excessive exposure to industrial chemicals, *Br. J. Ind. Med.*, 30, 289, 1973.
158. **Ogata, M., Takatsuka, Y., and Tomokuni, K.**, A simple method for the quantitative analysis of urinary trichloroethanol and trichloroacetic acid as an index of trichloroethylene exposure, *Br. J. Ind. Med.*, 27, 378, 1970.
159. **Daniel, J. W.**, The metabolism of ^{36}Cl-labelled trichloroethylene and tetrachloroethylene in the rat, *Biochem. Pharmacol.*, 12, 795, 1963.
160. **Ogata, M. and Saeki, T.**, Measurement of chloral hydrate, trichloroethanol, trichloroacetic acid and monochloroacetic acid in the serum and the urine by gas chromatography, *Int. Arch. Arbeitsmed.*, 33, 49, 1974.
161. **Mikisková, H. and Mikiska, A.**, Trichloroethanol in trichloroethylene poisoning, *Br. J. Ind. Med.*, 23, 116, 1966.
162. **Bartoníček, V. and Souček, B.**, Der Metabolismus des Trichloräthylens beim Kaninchen, *Arch. Gewerbepathol. Gewerbehyg.*, 17, 283, 1959.
163. **Stewart, R. D., Sadek, S. E., Swank, J. D., and Dodd, H. C.**, Diagnosis of trichloroethylene exposure after death, *Arch. Pathol.*, 77, 101, 1964.
164. **Byington, K. H. and Leibman, K. C.**, Metabolism of trichloroethylene in liver microsomes. II. Identification of the reaction product as chloral hydrate, *Mol. Pharmacol.*, 1, 247, 1965.
165. **Leibman, K. C. and McAllister, W. J., Jr.**, Metabolism of trichloroethylene in liver microsomes. III. Induction of the enzymic activity and its effect on excretion of metabolites, *J. Pharmacol. Exp. Ther.*, 157, 574, 1967.
166. **Ikeda, M.**, Reciprocal metabolic inhibition of toluene and trichloroethylene *in vivo* and *in vitro, Int. Arch. Arbeitsmed.*, 33, 125, 1974.
167. **Urban, Th. and Müller, G.**, Metabolization of trichloroethylene and its metabolites in rat liver, *Naunyn Schmiedebergs Arch. Pharmacol.*, 282, Suppl. R 100, 1974.
168. **Leibman, K. C.**, Metabolism of trichloroethylene in liver microsomes. I. Characteristics of the reaction, *Mol. Pharmacol.*, 1, 239, 1965.
169. **Kimmerle, G. and Eben, A.**, Metabolism, excretion and toxicology of trichloroethylene after inhalation. I. Experimental exposure on rats, *Arch. Toxicol.*, 30, 115, 1973.
170. **Ikeda, M. and Ohtsuji, H.**, A comparative study of the excretion of Fujiwara reaction-positive substances in urine of humans and rodents given trichloro- or tetrachloro-derivatives of ethane and ethylene, *Br. J. Ind. Med.*, 29, 99, 1972.
171. **Ikeda, M., Ohtsuji, H., Imamura, T., and Komoike, Y.**, Urinary excretion of total trichloro-compounds, trichloroethanol, and trichloroacetic acid as a measure of exposure to trichloroethylene and tetrachloroethylene, *Br. J. Ind. Med.*, 29, 328, 1972.
172. **Smyth, H. F., Jr., Carpenter, C. P., Weil, C. S., Pozzani, U. C., Striegel, J. A., and Nycum, J. S.**, Range-finding toxicity data: List VII, *Am. Ind. Hyg. Assoc. J.*, 30, 470, 1969.
173. **Christensen, H. E., Luginbyhl, T. T., and Carroll, B. S.**, The Toxic Substances List 1974 Edition, U.S. Department of Health, Education, and Welfare, Public Health Service Center for Disease Control, National Institute for Occupational Safety and Health, 1974, 353.

174. **Klaassen, C. D. and Plaa, G. L.,** Relative effects of various chlorinated hydrocarbons on liver and kidney function in mice, *Toxicol. Appl. Pharmacol.,* 9, 139, 1966.
175. **Schumacher, H. and Grandjean, E.,** Vergleichende Untersuchungen über die narkotische Wirksamkeit und die akute Toxicität von neun Losungsmitteln, *Arch. Gerwerbepathol. Gewerbehyg.,* 18, 109, 1960.
176. **Lazarew, N. W.,** Über die narkotische Wirkungskraft der Dämpfe der Chlorderivaten des Methans, des Äthans und des Äthylens, *Arch. Exp. Path. Pharmakol.,* 141, 19, 1929.
177. **Cresutelli, Y.,** Thesis, Würzburg, 1933. Cited by W. F. von Oettingen, *The halogenated hydrocarbons of industrial and toxicological importance,* U.S. Public Health Service, Elsevier, Amsterdam, 1964.
178. **Gehring, P. J.,** Hepatotoxic potency of various chlorinated hydrocarbon vapours relative to their narcotic and lethal potencies in mice, *Toxicol. Appl. Pharmacol.,* 13, 287, 1968.
179. **Carpenter, C. P., Smyth, H. F., Jr., and Pozzani, U. C.,** The assay of acute vapor toxicity, and the grading and interpretation of results on 96 chemical compounds, *J. Ind. Hyg. Toxicol.,* 31, 343, 1949.
180. **Siegel, J., Jones, R. A., Coon, R. A., and Lyon, J. P.,** Effects on experimental animals of acute, repeated and continuous inhalation exposures to dichloracetylene mixtures, *Toxicol. Appl. Pharmacol.,* 18, 168, 1971.
181. **Adams, E. M., Spencer, H. C., Rowe, V. K., McCollister, D. D., and Irish, D. D.,** Vapor toxicity of trichloroethylene determined by experiments on laboratory animals, *Arch. Ind. Hyg. Occup. Med.,* 4, 469, 1951.
182. **Prendergast, J. A., Jones, R. A., Jenkins, L. J., Jr., and Siegel, J.,** Effects on experimental animals of long-term inhalation of trichloroethylene, carbon tetrachloride, 1,1,1-trichlorethane, dichlorodifluoromethane, and 1,1-1-dichloroethylene, *Toxicol. Appl. Pharmacol.,* 10, 270, 1967.
183. **McCord, C. P.,** Toxicity of trichloroethylene, *J. Am. Med. Assoc.,* 99, 409, 1932.
184. **Matruchot, D.,** A comparison of the toxicity of the principal industrial solvents, *Presse Med.,* 47, 167, 1939.
185. **Aschieri, G., Majeron, M. A., and Toselli, E.,** Delirium Tremens da tricloroetilene (trielina), *Riv. Sper. Freniatr. Med. Leg. Alienazioni Ment.,* 92, 1516, 1968.
186. **Sutton, W. L.,** Psychiatric disorders and industrial toxicology, *Int. Psychiat. Clin.,* 6, 339, 1969.
187. **Zenk, H.,** Beruflich verursachte Vestibularisschäden, *Z. Aerztl. Fortbild.,* 64, 676, 1970.
188. **Illić, C., Nikloić, M., and Volćkov, V.,** Toksicne materije i kohleovestibularni aparat, *Med. Glas.,* 26, 56, 1972.
189. **Tomasini, M. and Sartorelli, E.,** Intossicazione cronica da inhalazione di trielina commerciale con compromissione dell'VII paio di nervi cranici, *Med. Lav.,* 62, 277, 1971.
190. **Tabacchi, G., Corsico, R., and Gallinelli, R.,** Neurite retrobulbare da sospetta intossicazione cronica da tricloroetilene, *Ann. Ottalmol. Clin. Ocul.,* 92, 787, 1966.
191. **Czerniawska, J.,** Wplyw trójchloroetylenu na narzad wzroku, *Klin. Oczna.,* 37, 123, 1967.
192. **Buxton, P. H. and Hayward, M.,** Polyneuritis cranialis associated with industrial trichloroethylene poisoning, *J. Neurol. Neurosurg. Psychiatry,* 30, 511, 1967.
193. **Feldman, R. G. and Mayer, R. F.,** Studies of trichloroethylene intoxication in man, *Neurology,* 18, 309, 1968.
194. **Feldman, R. G., Mayer, R. M., and Taub, A.,** Evidence for peripheral neurotoxic effect of trichloroethylene, *Neurology,* 20, 599, 1970.
195. **Mitchell, A. B. S. and Parsons-Smith, B. G.,** Trichloroethylene neuropathy, *Br. Med. J.,* 1, 422, 1969.
196. **Alapin, B. and Kozlowski, P.,** Narkomania trojchloroethylenowa u osobnika z wczesnym zanikiem mozgu, *Neurol. Neurochir. Psychiat. Pol.,* 10, 511, 1960.
197. **Kossakiewicz-Sulkonska, B.,** Przypadek zatrucia "Tri", przebiegajacego pod postacia zapalenia mozgu, *Pol. Tyg. Lek.,* 26, 1947, 1971.
198. **Sagawa, K., Nishitani, H., Kawai, H., Kuge, Y., and Ikeda, M.,** Transverse lesion of spinal cord after accidental exposure to trichloroethylene, *Int. Arch. Arbeitsmed.,* 31, 198 257, 1973.
199. **Bartoníček, V. J. and Brun, A.,** Subacute and chronic trichloroethylene poisoning: A neuropathological study in rabbits, *Acta. Pharmacol. Toxicol.,* 28, 359, 1970.
200. **Robiner, I. S.,** Clinico-electroencephalographic studies of trilene anesthesia, *Khirurgiya* (Moscow), 41, 81, 1965 (in Russian).
201. **Chalupa, B., Synková, J., and Sevčík, M.,** The assessment of electroencephalographic changes and memory disturbances in acute intoxications with industrial poisons, *Br. J. Ind. Med.,* 17, 238, 1960.
202. **Picotti, G. and Brugnone, F.,** Aspetti elettroencefalografici nell'intossicazione da trielina, *Folia Med.* (Naples), 45, 993, 1962.
203. **Roth, B. and Klimková-Deutschová, E.,** The effect of the chronic action of industrial poisons on the electroencephalogram of man, *Rev. Czech. Med.,* 9, 217, 1963.
204. **Roth, B. and Klimková-Deutschová, E.,** O účinku chronického působení průmyslových jedů na elektroencefalogram člověka, *Cesk. Neurol.,* 27, 40, 1964.
205. **Schwartzová, K.,** Nálezy EEG u chronické otravy trichlóretylénem, *Cesk. Neurol.,* 29, 369, 1966.
206. **Mellerio, F.,** Modifications électroencéphalographiques au cours d'intoxications aiguës par trichloréthylène, *Rev. Neurol.,* 121, 363, 1969.
207. **Mellerio, F.,** EEG changes during acute intoxication with trichlorethylene, *Electroencephalogr. Clin. Neurophysiol.,* 29, 101, 1970.
208. **Mellerio, F., Gaultier, M., Fournier, E., Gervais, P., and Frejaville, J.-P.,** Contribution of electroencephalography to resuscitation in toxicology, *Clin. Toxicol.,* 6, 271, 1973.

209. **Konietzko, H., Elster, I., Vetter, K., and Weichardt, H.,** Felduntersuchungen in Lösungsmittelbetrieben. 2. Telemetrische EEG- Überwachung bei Lösungmittelarbeitern an Trichlorathylen-Waschanlagen, *Z. Arbeitsmed. Arbeitsschutz,* 23, 129, 1973.
210. **Kurp, F. and Szymański, A.,** Zastosowanie metody analizy czynnikowej do oceny elektroencefalogramow, *Neurol. Neurochir. Pol.,* 7, 17, 1973.
211. **Mikiska, A. and Mikisková, H.,** Determining neurotoxicity by some electrographic methods (EEG, ECG, EMG) in guinea pigs, *Act. Nerv. Super.,* 6, 56, 1964.
212. **Horvath, M., Frantik, E., Korinek, F., Mikiska, A., and Mikiskova, H.,** Study of the function of the central nervous system in toxicologic experiments, *Gig. Tr. Prof. Zabol.,* 9, 9, 1965 (in Russian).
213. **Richards, C. D.,** Does trichlorethylene have a different mode of action from other general anaesthetics? *J. Physiol.,* 233, 25P, 1973.
214. **Angel, A., Berridge, D. A., and Unwin, J.,** The effect of anaesthetic agents on primary cortical evoked responses, *Br. J. Anaesth.,* 45, 824, 1973.
215. **Wilkinson, H. A., Mark, V. H., Wilson, R., and Patel, P.,** The toxicity of general anesthetics diffused directly into the brain, *Anesthesiology,* 38, 478, 1973.
216. **Mikisková, H.,** Srovnání účinku trichloretylénu a trichloretanolu na korové funkce a míšní reflexy, *Act. Nerv. Super.,* 4, 181, 1962.
217. **Robiner, I. S. and Svadzhyan, E. P.,** Changes in the bioelectric activity of the cortex and the subcortical structures of the brain in trilene anesthesia, *Eksp. Khir. Anesteziol.,* 13, 64, 1968 (in Russian).
218. **Heim, F., Estler, C.-J., Ammon, H. P. T., and Picker, D.,** Der Einfluss von Trichloräthylen auf den Gehalt des Gehirns weisser Mäuse an energiereichen Phosphaten, Coenzym A und Metaboliten des glykolytischen Kohlenhydratabbau, *Arch. Int. Pharmacodyn. Ther.,* 162, 311, 1966.
219. **Ungar, G.,** Inhibition of a brain protease by general anaesthetics, *Nature,* 207, 419, 1965.
220. **Grandjean, E.,** Trichloroethylene effects on animal behavior, *Arch. Environ. Health,* 1, 106, 1960.
221. **Zahner, H., Bättig, K., and Grandjean, E.,** Die Wirkung von Trichloraethylen auf das spontane links-rechts Alternieren der Ratte, *Med. Exp.,* 4, 191, 1961.
222. **Grandjean, E.,** Effets d'un narcotique sur le comportement conditionné et spontané du rat, *J. Physiol.* (Paris), 53, 353, 1961.
223. **Grandjean, E.,** Die Wirkung von Trichloräthylen auf das Verhalten von Ratten, *Pharm. Acta. Helv.,* 38, 464, 1963.
224. **Bättig, K. and Grandjean, E.,** Chronic effects of trichloroethylene on rat behavior, *Arch. Environ. Health,* 7, 694, 1963.
225. **Grandjean, E.,** The effects of short exposures to trichloroethylene on swimming performances and motor activity of rats, *Am. Ind. Hyg. Assoc. J.,* 24, 376, 1963.
226. **Baetjer, A. M., Annau, Z., and Abbey, H.,** Water deprivation and trichloroethylene, *Arch. Environ. Health,* 20, 712, 1970.
227. **Goldberg, M. E., Johnson, H. E., Pozzani, U. C., and Smyth, H. F., Jr.,** Behavioural response of rats during inhalation of trichloroethylene and carbon disulphide vapours, *Acta Pharmacol. Toxicol.,* 21, 36, 1964.
228. **Goldberg, M. E., Johnson, H. E., Pozzani, U.C., and Smyth, H. F., Jr.,** Effect of repeated inhalation of vapors of industrial solvents on animal behavior. I. Evaluation of nine solvent vapors on pole-climb performance in rats, *Am. Ind. Hyg. Assoc. J.,* 25, 369, 1964.
229. **Mikisková, H.,** Beitrag zur Verwendung der Abwehrreaktion auf die elektrische Hautreizung in der Toxikologie der Stoffe mit zentralnervoser Wirkung. I. Mitteilung Bestimmung der Latenz des Beugereflexes bei den Meerschweinchen, *Int. Arch. Gewerbepathol. Gewerbehyg.,* 19, 51, 1962.
230. **Mikisková, H. and Mikiska, A.,** Beitrag zur Verwendung der Abwehrreaktion auf die elektrische Hautreizung in der Toxikologie der Stoffe mit zentralnervöser Wirkung. II. Mitteilung Bestimmung der Reizschwellenintensität bei den Meerschweinehen, *Int. Arch. Gewerbepathol. Gewerbehyg.,* 19, 68, 1962.
231. **Mikisková, H. and Mikiska, A.,** Bestimmung der elektrischen Erregbarkeit der motorischen Grosshirnrinde und ihre Verwendung in der Pharmakologie und Toxikologie. III. Mitteilung Vergleish der narkotischen Wirkung von Trichloräthylen und Trichloräthanol bei den Meerschweinchen, *Int. Arch. Gewerbepath. Gewerbehyg.,* 18, 310, 1960.
232. **Horvath, M. and Formanek, J.,** Effect of small concentrations of trichloroethylene on the higher nervous activity in rats in chronic experimental conditions, *Zh. Vyssh. Nervn. Deyat. im. I. P. Pavlova.,* 9, 916, 1959 (in Russian).
233. **Defalque, R. J.,** The "specific" analgesic effect of trichloroethylene upon the trigeminal nerve, *Anesthesiology,* 22, 379, 1961.
234. **Neal, M. J. and Robson, J. M.,** The analgesic properties of sub-anaesthetic doses of anaesthetics in the mouse, *Br. J. Pharmacol.,* 22, 596, 1964.
235. **Neal, M. J. and Murray, B. R. P.,** The analgesic effect of anaesthetic mixtures. The effect of nitrous oxide with trichloroethylene or halothane on experimental ischaemic pain, *Guy's Hosp. Rep.,* 115, 19, 1966.
236. **Mroczek, H. and Fedyk, T.,** Przpadek ciezkiego, samobójczego zatrucia trójchloroetylenem "Tri", *Pol. Tyg. Lek.,* 26, 1509, 1971.
237. **Kledecki, Z. and Bura, H.,** Ostre zatrucia wywolane polknieciem trójchloroetylenu ("Tri"), *Pol. Tyg. Lek.,* 18, 748, 1963.

238. **Yacoub, M., Faure, J., Rollux, R., Mallion, J.-M., and Marka, C.**, L'intoxication aiguë par le trichloréthylène. Manifestations cliniques et paracliniques, surveillance de l'élimination du produit, *J. Eur. Toxicol.*, 6, 275, 1973.
239. **Konietzko, H. and Elster, I.**, Cardiotoxische Wirkungen von Trichloräthylen, *Arch. Toxikol.*, 34, 93, 1973.
240. **Scorsone, A., Coppola, A., and Rizzo, A.**, Rilievi eegrafici in soggetti esposti al rischio di tricloroetilene, *Boll. Soc. Ital. Cardiol.*, 11, 762, 1966.
241. **Pebay-Peyroula, F., Le Gall, J. R., Fréjaville, J. P., Rosenthal, D., and Gaultier, M.**, 69 cas d'intoxication aiguë par le trichloréthylène, *Bull. Soc. Med. Hosp. Paris.*, 117, 1137, 1966.
242. **Dervillee, P.**, Intossicazione da Trichloroetilene, *Folia Med.* (Naples), 47, 105, 1964.
243. **Lilis, R., Stanescu, D., Muica, N., and Roventa, A.**, Chronic effects of trichloroethylene exposure, *Med. Lav.*, 60, 595, 1969.
244. **Dimitrova, M., Usheva, G., and Pavlova, S.**, The work environment's influence on the cardiovascular system. Polycardiographic investigations in workers exposed to trichloroethylene, *Int. Arch. Arbeitsmed.*, 32, 145, 1974.
245. **Kiessling, W.**, Vergleichende Untersuchungen uber die Wirkung einiger Chlorderivate des Methans, Äthans und Äthylens am isolierten Froschherzen, *Biochem. Z.*, 114, 292, 1921.
246. **Bianchi, A., DeNatale, G., and Matturro, F.**, Ricerche comparative su alcuni effetti farmacologici del Fluothane e di altri narcotici volatili, *G. Ital. Chir.*, 19, 327, 1963.
247. **Matturro, F.**, Effetti di alcuni narcotici volatili sul cuore isolato di cavia, *G. Ital. Chir.*, 19, 95, 1963.
248. **Reinhardt, C. F., Mullin, L. S., and Maxfield, M. E.**, Epinephrine-induced cardiac arrhythmia potential of some common industrial solvents, *J. Occup. Med.*, 15, 953, 1973.
249. **Evreux, J. C., Motin, J., Vincent, V., and Faucon, G.**, Essai du propanolol dans les troubles du rythme cardiaque d'intoxications expérimentales par le trichloréthylène, *Med. Pharmacol. Exp.*, 17, 527, 1967.
250. **Murray, W. J., McKnight, R. L., and Davis, D. A.**, Antagonism of hydrocarbon anesthetic-epinephrine arrhythmias in dogs by Nethalide, a dichloroisoproterenol analogue, *Proc. Soc. Exp. Biol. Med.*, 113, 439, 1963.
251. **Evreux, J.-C., Ducluzeau, R., Vincent, V., Arcadio, F., and Motin, J.**, Les troubles cardiaques au cours des intoxications aiqués par le trichloréthylène, Etude expérimentale de leur traitement, *Lyon Med.*, 219, 167, 1968.
252. **Orlando, E., Raffi, G. B., and Alessandri, M.**, Turbe del ritmo cardiaco in osservazione sperimentale di intossicazione acute da trielina, *G. Clin. Med.* (Bologna), 47, 162, 1966.
253. **Fusco, M. and Brancaccio, A.**, Il vettorcardiogramma Nell'intossicazione subacuta sperimentale da trichloroetilene, *Folia Med.* (Naples), 51, 362, 1968.
254. **Matteo, R. S., Katz, R. L., and Papper, E. M.**, The injection of epinephrine during general anesthesia with halogenated hydrocarbons and cyclopropane in man. 1. Trichoroethylene, *Anesthesiology*, 23, 360, 1962.
255. **Wilhjelm, B. J. and Arnfred, I.**, Protective action of some anaesthetics against anoxia, *Acta. Pharmacol. Toxicol.*, 22, 93, 1965.
256. **Nisbet, H. I. A., Gray, I. G., Olley, P. M., and Johnston, A. E.**, Cardiovascular and respiratory responses to severe hypoxaemia during anaesthesia. 1. The effect of various concentrations of three anaesthetic agents upon the cardiovascular response and oxygen transport, *Can. Anaesth. Soc. J.*, 19, 339, 1972.
257. **Dobkin, A. B., Byles, P. H., and Neville, J. F., Jr.**, Neuroendocrine and metabolic effects of general anaesthesia and graded haemorrhage, *Can. Anaesth. Soc. J.*, 13, 453, 1966.
258. **Smith, G. and Ledingham, I. McA.**, The effect of prolonged hyperoxia on the cardiovascular system of anaesthetized dogs, *Br. J. Anaesth.*, 44, 469, 1972.
259. **Ellis, F. R.**, Blood loss from conjunctival wounds: a comparison of halothane and trichloroethylene, *Br. J. Anaesth.*, 38, 941, 1966.
260. **McDowall, D. G., Barker, J., and Jennett, W. B.**, Cerebro-spinal fluid pressure measurements during anaesthesia, *Anaesthesia*, 21, 189, 1966.
261. **McDowall, D. G., Harper, A. M., and Jacobson, I.**, Cerebral blood flow during trichloroethylene anaesthesia: a comparison with halothane, *Br. J. Anaesth.*, 36, 11, 1964.
262. **McDowall, D. G.**, The effects of general anaesthetics on cerebral blood flow and cerebral metabolism, *Br. J. Anaesth.*, 37, 236, 1965.
263. **McDowall, D. G. and Harper, A. M.**, Blood flow and oxygen uptake of the cerebral cortex of the dog during anaesthesia with different volatile agents, *Acta. Neurol. Scand.*, 41 (Suppl. 14), 146, 1965.
264. **Kékesi, F., Gallyas, F., and Szántó, J.**, Die Wirkung von verschiedenen Inhalationsnarkosetypen auf Die Gehirndurchblutung, *Acta. Med. Acad. Sci. Hung.*, 24, 153, 1967.
265. **Mazza, A. and Munarini, D.**, Su di un caso di vasculopatia cerebrale da intossicazione da tricloroetilene, *Riv. Sper. Freniatr.*, 84, 407, 1960.
266. **McArdle, L., Unni, V. K. N., and Black, G. W.**, The effects of trichloroethylene on limb blood flow in man, *Br. J. Anaesth.*, 40, 767, 1968.
267. **Edwards, J. C. and Fuzzey, G. J. J.**, The effect of anaesthetic agents upon calf muscle blood flow in the ischaemic limb, *Br. J. Anaesth.*, 42, 514, 1970.
268. **Unni, V. K. N., McArdle, L., and Dundee, J. W.**, Peripheral vascular effects of pethidine and pentazocine during trichloroethylene anaesthesia, *Br. J. Anaesth.*, 44, 593, 1972.
269. **Omarova, G. V.**, Indices of blood supply and tonus of cerebral vessels in dental patients under rotilan (trichlorethylene) analgesia according to rheoencephalographic data, *Stomatologia*, 50, 24, 1971 (in Russian).

270. **Danilenko, M. V. and Timchuk, I. D.**, Hemodynamic changes in major and minor surgery carried out under trichloroethyl anesthesia, *Klin. Khir.*, 6, 42, 1968 (in Russian).
271. **Damir, E. A., Gulyaev, G. V., Solomonik, V. Z., and Tatarsky, M. L.**, Combined trichloroethylene anesthesia in neurosurgery (clinical study of the anesthesia and of changes occurring in the vascular tone), *Vopr. Neirokhir.*, 29, 36, 1965 (in Russian).
272. **Meyer, H. J.**, Altérations bronchopulmonaires par le trichloréthylène et autres hydrocarbures halogenés, *Bronches*, 23, 113, 1973.
273. **Jouglard, J. and Vincent, V.**, Les indices pulmonaires des ingestions de solvants, *Marseille Med.*, 108, 696, 1971.
274. **Consorti, P. and Giomarelli, P. P.**, Indagine sull'eventulale azione battericida o batteriostatica dell'etere etilico, tricloroetilene, fluotano, metossifluorano, *Atti. Accad. Fisiocrit. Siena Sez. Med. Fis.*, 17, 128, 1968.
275. **Mehta, S., Behr, G., and Kenyon, D.**, The effect of volatile anaesthetics on common respiratory pathogens: halothane, trichloroethylene and methoxyflurane, *Anaesthesia*, 29, 280, 1974.
276. **Mehta, S., Behr, G., and Kenyon, D.**, The effect of volatile anaesthetics on bacterial growth, *Can. Anaesth. Soc. J.*, 20, 230, 1973.
277. **Dobkin, A. B., Byles, P. H., and Neville, J. F., Jr.**, Neuroendocrine and metabolic effects of general anaesthesia during spontaneous breathing, controlled breathing, mild hypoxia, and hypercarbia, *Can. Anaesth. Soc. J.*, 13, 130, 1966.
278. **Malchy, H. and Parkhouse, J.**, Respiratory studies with trichloroethylene, *Can. Anaesth. Soc. J.*, 16, 119, 1969.
279. **Unni, V. K. N., McArdle, L., and Black, G. W.**, Sympatho-adrenal, respiratory and metabolic changes during trichloroethylene anaesthesia, *Br. J. Anaesth.*, 42, 429, 1970.
280. **Unni, V. K. N., McArdle, L., and Dundee, J. W.**, Respiratory effects of pethidine and pentazocine during trichloroethylene anaesthesia, *Br. J. Anaesth.*, 44, 692, 1972.
281. **Unni, V. K. N., McArdle, L., and Dundee, J. W.**, Respiratory effects of pethidine and pentazocine during trichloroethylene anaesthesia (letter), *Br. J. Anaesth.*, 45, 85, 1973.
282. **Dobkin, A. B., Harland, J. H., and Fedoruk, S.**, Trichlorethylene and halothane in a precision system: comparison of cardiorespiratory and metabolic effects in dogs, *Anesthesiology*, 23, 58, 1962.
283. **Loh, L., Seed, R. F., and Sykes, M. K.**, The cardiorespiratory effects of halothane, trichloroethylene and nitrous oxide in the dog, *Br. J. Anaesth.*, 45, 125, 1973.
284. **Kubacki, A., Mrozikiewicz, A., Skubiszyńska, A., and Wachowiak, A.**, Wplyw vandidu na czynność serca w doświadczalnym zatruciu trójchloroetylenem, *Med. Pr.*, 17, 196, 1966.
285. **Ngai, S. H., Katz, R. L., and Farhie, S. E.**, Respiratory effects of trichlorethylene, halothane and methoxyflurane in the cat, *J. Pharmacol. Exp. Ther.*, 148, 123, 1965.
286. **Harrison, G. A., Moir, D. D., and Vanik, P. E.**, The sensitivity of the respiratory tract during anaesthesia in the cat, *Br. J. Anaesth.*, 35, 403, 1963.
287. **Coleridge, H. M., Coleridge, J. C. G., Luck, J. C., and Norman, J.**, The effect of four volatile anaesthetic agents on the impulse activity of two types of pulmonary receptor, *Br. J. Anaesth.*, 40, 484, 1968.
288. **Sykes, M. K., Davies, D. M., Chakrabarti, M. K., and Loh, L.**, The effects of halothane, trichloroethylene and ether on the hypoxic pressor response and pulmonary vascular resistance in the isolated, perfused cat lung, *Br. J. Anaesth.*, 45, 655, 1973.
289. **Parish, W. E., Hall, L. W., and Coombs, R. R. A.**, The effect of anaesthesia on anaphylaxis in guinea-pigs, *Immunology*, 6, 462, 1963.
290. **Ghosh, M. N., Srivastava, R. K., and Ghosh, A. K.**, Safety index of ether, chloroform, trichloroethylene and halothane in mice, rats and guinea pigs, *Arch. Int. Pharmacodyn. Ther.*, 138, 548, 1962.
291. **Ghosh, M. N., Srivastava, R. K., Ghosh, A. K.**, Dose-response relationships of ether, chloroform, trichloroethylene and halothane in mice and rats, *Arch. Int. Pharmacodyn. Ther.*, 150, 96, 1964.
292. **Truhaut, R., Boudène, C., Jouany, J-M., and Bouant, A.**, Application du physiogramme à l'étude de la toxicologie aiguë des solvants chlorés, *Eur. J. Toxicol. Hyg. Environ.*, 5, 284, 1972.
293. **Priest, R. J. and Horn, R. C., Jr.**, Trichloroethylene intoxication, a case of acute hepatic necrosis possibly due to this agent, *Arch. Environ. Health*, 11, 361, 1965.
294. **Ossenberg, F. W., Martin, W., Saegler, J., and Hann, D.**, Le variazioni delle attivita enzimatiche durante le intossicazioni, *Minerva Med.*, 63, 3027, 1972.
295. **Chiesura, P. and Corsi, G.**, Intossicazione acuta umana da tricloroetilene seguita da epatopatia e da glicosuria iperglicemica, *Folia Med.* (Naples), 44, 121, 1961.
296. **Albahary, C., Guyotjeannin, C., Flaisler, A., and Thiaucourt, P.**, Transaminases et exposition professionnelle au trichloréthylène, *Arch. Mal. Prof. Med. Trav. Secur. Soc.*, 20, 421, 1959.
297. **Schüttmann, W.**, Zur Frage der Leberschädigung durch beruflichen Kontakt mit Trichloräthylen, *Dtsch. Z. Verdau. Stoffwechselkr.*, 30, 43, 1970.
298. **Lachnit, V.**, Halogenisierte Kohlenwasserstoffe und Leber, *Wien Klin. Wochenschr.*, 83, 723, 1971.
299. **Szadkowski, D. and Körber, M.**, Leberfunktionsprüfungen bei Lösemittel-exponierten Werktätigen in der metallverarbeitenden Industrie, *Int. Arch. Gewerbepathol. Gewerbehyg.*, 25, 323, 1969.
300. **Tolot, F., Viallier, J., Roullet, A., Rivoire, J., and Figueres, J. C.**, Toxicité hépatique du trichloréthylène, *Arch. Mal. Prof. Med. Trav. Secur. Soc.*, 25, 9, 1964.

301. **Šarić, M., Prpić-Majić, D., and Čudina-Nikšić, Z.,** Aktivnost serumske transaminaze kod profesionalne ekspozicije trikloretilenu, *Arh. Hig. Rada Toksikol.*, 13, 183, 1962.
302. **Karavanov, G. G. and Borzhiyevsky, C. K.,** Comparative evaluation of the effect of different types of anesthesia on the liver as judged by sorbitdehydrogenase activity, *Vrach. Delo.*, 6, 23, 1972 (in Russian).
303. **Bløndal, B. and Fagerlund, B.,** Trichlorethylene anaesthesia and hepatic function, *Acta Anaesthesiol. Scand.*, 7, 147, 1963.
304. **Kylin, B., Reichard, H., Sümegi, and Yllner, S.,** Hepatotoxic effect of tri- and tetra-chlorethylene on mice, *Nature*, 193, 395, 1962.
305. **Kylin, B., Reichard, H., Sümegi, and Yllner, S.,** Hepatotoxicity of inhaled trichloroethylene, tetrachloroethylene and chloroform. Single exposure, *Acta Pharmacol. Toxicol.*, 20, 16, 1963.
306. **Ikeda, T., Hagano, C., and Okada, A.,** Hepatotoxic effect of trichloroethylene and perchloroethylene in rats and mice, *Med. Biol.* (Tokyo), 79, 123, 1969 (in Japanese).
307. **Heim, F., Estler, C-J., Ammon, H. P. T., and Hähnel, U.,** Der Metabolitgehalt der Leber weisser Mäuse bei akuter Trichloräthylenvergiftung, *Med. Pharmacol. Exp.*, 15, 116, 1966.
308. **Zadorozhnyi, B. V.,** Changes in the body of animals during prolonged inhalation exposure to trichloroethylene in low concentrations, *Gig. Tr. Prof. Zabol.*, 17, 55, 1973 (in Russian).
309. **Mackiewicz, U., Pech, J., and Stasińska, M.,** Wplyw narkozy chloroformoweij, trójchloroetylenowej i fluotanowej na czynność watroby królika, *Acta Pol. Pharm.*, 23, 383, 1966.
310. **Corsi, G. E., Galzigna, L., and Brugnone, F.,** Azione del trichloroetilene su una via catabolica del triptofano, *Folia Med.* (Naples), 46, 858, 1963.
311. **Klaassen, C. D. and Plaa, G. L.,** Relative effects of various chlorinated hydrocarbons on liver and kidney function in dogs, *Toxicol. Appl. Pharmacol.*, 10, 119, 1967.
312. **Verne, J., Ceccaldi, P-F., Hébert, S., and Roux, J. M.,** Recherches sur la stéatose hépatique au cours des intoxications par des composés organiques volatils. IV. Etudes biochimique et histochimique des foies gras obtenus par intoxication au trichloréthylène, *Pathol. Biol.*, 7, 2311, 1959.
313. **Verne, J., Hébert, S., and Roux, J.,** Les reactions de la cellule hepatique au cours des intoxications experimentales par des composes organiques volatils chlores, *An. Fac. Med. (Montevideo)*, 44, 476, 1959.
314. **Tronche, P., Bressolette, M. H., and Laroussinie, C.,** Composition de graisses phosphorées du foie de rat dans l'intoxication expérimentale au trichloréthylène, *C. R. Soc. Biol.*, 155, 1508, 1961.
315. **Wirtschafter, Z. T. and Cronyn, M. W.,** Relative hepatotoxicity: pentane, trichloroethylene, benzene, carbon tetrachloride, *Arch. Environ. Health*, 9, 180, 1964.
316. **Johnson, M. K.,** The influence of some aliphatic compounds on rat liver glutathione levels, *Biochem. Pharmacol.*, 14, 1383, 1965.
317. **Bloxam, D. L.,** Effects of various anaesthetics on the metabolism and general condition of the isolated perfused rat liver, *Biochem. Pharmacol.*, 16, 283, 1967.
318. **Ramadan, M. A. and Ramadan, M. I. A.,** Histochemical studies on the effect of some anaesthetics on rat liver, *Acta Histochem.*, 34, 310, 1969.
319. **Kiseleva, A. F. and Korolenko, A. M.,** Morpho-histochemical changes in the liver under the effect of anesthetic doses of phtorotane and trilen in the experiment, *Vrach. Delo.*, No. 1, 12, 1971 (in Russian).
320. **Korolenko, A. M.,** Electron-microscopic studies of liver changes due to phtorotane and trilen anesthesia, *Vrach. Delo.*, No. 6, 28, 1972 (in Russian).
321. **Borzhievski, T. K.,** Estimation of anesthetics according to their hepatotoxicity and ornithine carbamoyl-transferase activity, *Vestn. Khir. im. I. I. Grekova.*, 107, 97, 1971 (in Russian).
322. **Bloxam, D. L.,** Effects of halothane, trichloroethylene, pentobarbitone and thiopentone on amino acid transport in the perfused rat liver, *Biochem. Pharmacol.*, 16, 1848, 1967.
323. **Windorfer, A. and Stier, A.,** Physikalisch-chemische Faktoren bei der Bindung von Halogenkohlen-wasserstoffen an Lebermikrosomen, *Naunyn Schmiedebergs Arch. Pharmakol.*, 263, 258, 1969.
324. **Cornish, H. H., Ling, B. P., and Barth, M. L.,** Phenobarbital and organic solvent toxicity, *Am. Ind. Hyg. Assoc. J.*, 34, 487, 1973.
325. **Carlson, G. P.,** Enhancement of the hepatotoxicity of trichloroethylene by inducers of drug metabolism, *Res. Commun. Chem. Pathol. Pharmacol.*, 7, 637, 1974.
326. **Traiger, G. J. and Plaa, G. L.,** Chlorinated hydrocarbon toxicity. Potentiation by isopropyl alcohol and acetone, *Arch. Environ. Health*, 28, 276, 1974.
327. **Cornish, H. H. and Adefuin, J.,** Ethanol potentiation of halogenated aliphatic solvent toxicity, *Am. Ind. Hyg. Assoc. J.*, 27, 57, 1968.
328. **Desoille, H., Pinchon, R. A., Jans, M., and Bourguignon, A.,** Intoxication expérimentale aiguë par le trichloréthylène. Effets aggravants de l'intoxication éthylique chronique associée. Étude experimentale électroencéphalographique chez le lapin, *Arch. Mal. Prof. Med. Trav. Secur. Soc.*, 23, 653, 1962.
329. **Sbertoli, C. and Brambilla, G.,** Tre casi di intolleranza all'alcool come unico sintomo della intossicazione da tricloroetilene, *Med. Lav.*, 53, 353, 1967.
330. **Pardys, S. and Brotman, M.,** Trichloroethylene and alcohol: a straight flush (letter), *J. Am. Med. Assoc.*, 229, 521, 1974.

331. **Lob, M.**, L'action du trichloréthylène su le taux d'alcool dans le sang, *Med. Lav.*, 51, 587, 1960.
332. **Chiesura, P.**, Sulla nefropatia da tricloroetilene, *Minerva Nefrol.*, 6, 95, 1959.
333. **Gutch, C. F., Tomhave, W. G., and Stevens, S. C.**, Acute renal failure due to inhalation of trichlorethylene, *Ann. Intern. Med.*, 63, 128, 1965.
334. **Ponticelli, C., Imbasciati, E., and Redaelli, B.**, Su due casi di insufficienza renale acuta da tricloroetilene, *Minerva Nefrol.*, 15, 146, 1968.
335. **Pelka, W. and Zach, E.**, Ostra niewydolność nerek w przebiegu ostrego zatrucia trójchloroetylenem, *Wiad. Lek.*, 27, 539, 1974.
336. **Haug, E.**, Akutt trikloretylenforgiftning. Akutt nyresvikt etter trikloretylenin-halasjon behandlet med peritoneal dialyse, *Tidsskr. Nor. Laegeforen.*, 90, 288, 1970.
337. **Kiseleva, A. F. and Korolenko, A. M.**, Histoenzymatic changes in the liver and kidneys during the experimental action of anesthetic doses of fluorothane and trilene, *Eksp. Khir. Anesteziol.*, 16, 81, 1971 (in Russian).
338. **Pennarola, R., Lamanna, P., and Castellino, N.**, Reperti istopatologici nell'intossicazione sperimentale da tricloroetilene, *Folia Med.* (Naples), 49, 853, 1966.
339. **Mazza, V., Brancaccio, A., and Elmino, O.**, La funzione renale nell'intossicazione sperimentale da tricloroetilene, *Folia Med.* (Naples), 51, 282, 1968.
340. **Kadlec, K.**, Profesionální kontaktní ekzémy při práci s trichlóretylénem, *Cesk. Dermatol.*, 38, 395, 1963.
341. **Peck, S. M.**, Dermatitis from cutting oils, solvents and dielectrics, including chloracne, *J. Am. Med. Assoc.*, 125, 190, 1944.
342. **Schirren, J. M.**, Hautschädigungen durch Trichloraethylen (Tri) in einem metallverarbeitenden Betrieb, *Berufs-Dermatosen*, 19, 240, 1971.
343. **Bauer, M. and Rabens, S. F.**, Cutaneous manifestations of trichloroethylene toxicity, *Arch. Dermatol.*, 110, 886, 1974.
344. **Leong, B. K. J., Schwetz, B. A., and Gehring, P. J.**, Embryo- and fetotoxicity of inhaled trichloroethylene, perchloroethylene, methylchloroform and methylene chloride in mice and rats (abstract), Society of Toxicology, 14th Annual Meeting, Williamsburg, Virginia, March 9–13, 1975, 29.
345. **Euler, H. H.**, Tier experimentelle Untersuchung einer Industrie-Noxe, *Arch. Gynaekol.*, 204, 258, 1967.
346. **Laham, S.**, Studies on placental transfer. Trichlorethylene, *Industr. Med. Surg.*, 39, 46, 1970.
347. **Rooth, G.**, Influence of nitrous oxide on the acid-base balance of the cord blood, *Am. J. Obstet. Gynecol.*, 85, 48, 1963.
348. **Moya, F. and Thorndike, V.**, The effects of drugs used in labor on the fetus and newborn, *Clin. Pharmacol. Ther.*, 4, 623, 1963.
349. **Tsygankova, S. T. and Kastrubin, E. M.**, The state of gas exchange and acid-base balance of the fetus in labor anesthesia with a mixture of trilen and oxygen, *Akush. Ginekol.* (Moskow), 42, 17, 1966 (in Russian).
350. **Kalganova, R. I., Tsigankova, S. T., and Chernobai, V. P.**, Fetal indices of gaseous exchange and acid-alkaline balance in labour analgesia with trichlorethylene combined with oxygen or air, *Vopr. Okhr. Materin. Det.*, 14, 81, 1969 (in Russian).
351. **Korobko, S. F.**, The course of labour and the indices of the fetal blood acid-base balance in trichloroethylene and nitrous oxide anesthesia, *Vopr. Okhr. Materin. Det.*, 17, 56, 1972 (in Russian).
352. **Phillips, T. J. and MacDonald, R. R.**, Comparative effect of pethidine, trichloroethylene, and entonox on fetal and neonatal acid-base and PO_2, *Br. Med. J.*, 3, 559, 1971.
353. **Lilleaasen, P.**, Narkose og anestesi som årsak til fødselsasfyksi, *Tidsskr. Nor. Laegeforen.*, 92, 1938, 1972.
354. **Lakomy, T., Papierowski, Z., and Kylszejko, Cz.**, Résultats cliniques de l'influence du trichloréthylène (TCE) sur la contractilité de l'utérus gravide pendant l'accouchement comparés aux résultats des examens *in vitro, Gynecol. Obstet.*, 64, 665, 1972.
355. **Shanks, C. A.**, The compatibility of octapressin with cyclopropane, trichlorethylene and halothane, *N. Z. Med. J.*, 63, 156, 1964.
356. **Klyszejko, C., Lakomy, T., and Papierowski, Z.**, Wplyw trójchloroetylenu na czynność skurczowa macicy w czasie porodu w zestawieniu z wynikami in vitro, *Pol. Tyg. Lek.*, 18, 1333, 1963.
357. **Van Liere, E. J., Mazzocco, T. R., and Northup, D. W.**, The effect of cyclopropane, trichlorethylene, and ethyl chloride on the uterus of the dog, *Am. J. Obstet. Gynecol.*, 94, 861, 1966.
358. **Sternadel, Z.**, Trójchloroetylen jako analgetyk w poloznictwie (120 przypadków wlasnych), *Pol. Tyg. Lek.*, 16, 1191, 1961.
359. **Thierstein, S. T., Hanigan, J. J., Faul, M. D., and Stuck, P. L.**, Trichloroethylene anesthesia in obstetrics: report of 10,000 cases, with fetal mortality and electrocardiographic data, *Obstet. Gynecol.*, 15, 560, 1960.
360. **Krasovitskaya, M. L. and Malyarova, L. K.**, On the chronic effect of small concentrations of ethylene and trichloroethylene on the organism of newborn animals, *Gig. Sanit.*, 33, 7, 1968 (in Russian).
361. **Wink, A.**, Effect of long-term exposure to low levels of toxic substances on urinary excretion of 17-oxogenic steroids and 17-oxosteroids, *Ann. Occup. Hyg.*, 15, 211, 1972.
362. **Żrubek, H., Tynecki, J., Dabek, W., and Czekierdowska, D.**, Aktywność adrenooksydazy w zwiazku z zabiegiem przerywania ciazy w znieczuleniu trichlorenem i nowokaina, *Pol. Tyg. Lek.*, 21, 1481, 1966.
363. **Dmitrieva, N. V.**, Bioelectrical activity and electroconductive properties of the muscles under the action of chlorinated hydrocarbons, *Farmakol. Toksikol.* (Moscow), 31, 228, 1968 (in Russian).

364. **Basalaev, A. V.**, Experience with the use of large-frame photofluorography in examining skeletal bones of persons occupationally dealing with unsaturated hydrocarbons of the ethylene series (olefins) and their chlorine derivatives (vinyl chloride, trichlorethylene), *Gig. Tr. Prof. Zabol.*, 14, 340, 1970 (in Russian).
365. **Savić, S., Baaske, H., and Hockwin, O.**, Biochemische Veränderungenam Kaninchenauge bei Vergiftung mit Trichloräthylen, *Arch. Klin. Exp. Ophthalmol.*, 175, 1, 1968.
366. **Savić, S. and Hockwin, O.**, Biochemiche Veränderungenam Kaninchenauge bei Vergiftung mit organischen Lösungsmitteln, *Ophthalmologica*, 359, 63, 1969.
367. **Gerritsen, B. G.**, The effect of anaesthetics on the electroretinogram and the visually evoked response in the rabbit, *Doc. Ophthalmol.*, 29, 289, 1971.
368. **Oduntan, S. A. and Akinyemi, O. O.**, Post-operative vomiting in the African, *West Afr. Med. J.*, 19, 176, 1970.
369. **Gombos, F., D'Urso, M., and Marenduzzo, A.**, Occupational diseases of the oral cavity: 3. Effects of some poisons and dusts on the submandibular and parotid salivary glands on the extraorbital lacrimal gland of the albino rat: experimental research, *Riv. Ital. Stomatol.*, 27, 149, 1972.
370. **Speden, R. N.**, The effect of some volatile anaesthetics on the transmurally stimulated guinea-pig ileum, *Br. J. Pharmacol.*, 25, 104, 1965.
371. **Rang, H. P.**, Stimulant actions of volatile anaesthetics on smooth muscle, *Br. J. Pharmacol.*, 22, 356, 1964.
372. **Matturro, J. and De Natale, G.**, Effetti di alcuni narcotici volatili sull'intestino isolato di cavia, *G. Ital. Chir.*, 19, 205, 1963.
373. **Guyotjeannin, C. and Guyotjeannin, N.**, Studio della coagulazione sanguigna fra operai esposti a vapori di tricloroetilene, *Folia. Med.* (Naples), 42, 924, 1959.
374. **Pinkhas, J., Cohen, I., Kruglak, J., and de Vries, A.**, Hobby-induced factor VII deficiency, *Haemostasis*, 1, 52, 1972.
375. **Mazza, V. and Brancaccio, A.**, Comportamento degli elementi figurati del sangue e del midollo nella intossicazione sperimentale da Tri, *Folia Med.* (Naples), 50, 318, 1967.
376. **Coppola, A. and Rizzo, A.**, Sull'azione tossica del tricloroetilene, *Folia Med.* (Naples), 49, 376, 1966.
377. **Pelts, D. G.**, Effect of trichloroethylene and tetrachloroethylene on the phagocyte activity of the leukocytes of the blood, *Gig. Sanit.*, 27, 96, 1962 (in Russian).
378. **Friborská, A.**, The phosphatases of peripheral white blood cells in workers exposed to trichloroethylene and perchloroethylene, *Br. J. Ind. Med.*, 26, 159, 1969.
379. **Vaccaro, U.**, L'indice opsonico del siero di sangue nella intossicazione da tricloroetilene, *Minerva Medicoleg.*, 80, 203, 1960.
380. **Dobkin, A. B. and Byles, P. H.**, The effect of trichloroethylene-nitrous oxide anaesthesia on acid-base balance in man, *Br. J. Anaesth.*, 34, 797, 1962.
381. **Buchan, A. S. and Bauld, H. W.**, Blood-gas changes during trichloroethylene and intravenous pethidine anaesthesia, *Br. J. Anaesth.*, 45, 93, 1973.
382. **Shmuter, L. M.**, Mechanism of the effect of trichloroethylene poisoning on the formation of antibodies against the o-antigen of salmonella typhi, *Byull. Eksp. Biol. Med.*, 74, 77, 1972 (in Russian).
383. **Shmuter, L. M.**, Effect of chronic action of small concentrations of chlorinated hydrocarbons on the production of various classes of immunoglobulins, *Gig. Sanit.*, 37, 36, 1972 (in Russian).
384. **Shmuter, L. M.**, Mechanism of the effect of trichloroethylene poisoning on antibody formation to the o-antigen of salmonella typhi, *Bull. Exp. Biol. Med.*, 74, 945, 1972.
385. **Navrotskii, V. K., Kashin, L. M., Kulinskaya, I. L., Mikhailovskaya, L. F., Shmuter, L. M., Burlaka-Vovk, Z. I., and Zadorozhnyi, B. V.**, Comparative evaluation of the toxicity of a series of industrial poisons during their long-term inhalation action in low concentrations, *Tr. S'ezda. Gig. Ukr. Ssr.*, 8, 224, 1971 (in Russian).
386. **Holmberg, B. and Malmfors, T.**, The cytotoxicity of some organic solvents, *Environ. Res.*, 7, 183, 1974.
387. **Longley, E. O. and Jones, R.**, Acute trichloroethylene narcosis. Accident involving the use of trichloroethylene in a confined space, *Arch. Environ. Health*, 7, 249, 1963.
388. **Mouton, N.**, A propos de 28 cas d'intoxication aigue collective et accidentelle par trichloréthylène, *Arch. Mal. Prof. Med. Trav. Secur. Soc.*, 33, 66, 1972.
389. **Gaultier, M., Efthymiou, M. L., Efthymiou, Th., and Pebay-Peyroula, F.**, Manifestations cardiaques de l'intoxication par le trichloréthylène, *Ann. Cardiol. Angeiol.*, 20, 185, 1971.
390. **Jouglard, J., Ohresser, Ph., Gerin, J., Laurent, Y., Terrasson de Fougeres, P.**, L'intoxiciation aiguë par le trichloréthylène, *Marseille Med.*, 107, 879, 1970.
391. **Ducluzeau, R., Bouletreau, P, Motin, J., Vincent, V., Demiaux, J. P., and Rouzioux, J. M.**, Intoxications au trichloréthylène, *Lyon Med.*, 217, 1159, 1967.
392. **Roche, L., Saury, A., and Girard, R.**, Difficultés de diagnostic de l'intoxication benigne au trichloréthylène, *Arch. Mal. Prof. Med. Trav. Secur. Soc.*, 27, 700, 1966.
393. **Rendoing, J., Seys, G. A., Manceaux, J. C., and Choisy, H.**, Intoxication collective accidentelle par vapeurs de trichloréthylène, *Eur. J. Toxicol. Hyg. Environ.*, 6, 284, 1973.
394. **Gibitz, H. J. and Plöchl, E.**, Orale trichloräthylenvergiftung bei einem 4½ Jahre alten Kind, *Arch. Toxicol.*, 31, 13, 1973.
395. **Neuendorff, C.**, Uber erste Ergebnisse prophylaktischer Trichloräthylenuntersuchungen in einem Textilreinigungsbetrieb, *Z. Gesamte. Hyg. Ihre Grenzgeb.*, 13, 728, 1967.

396. **Sukhotina, K. I., Sukhanova, V. A., Dumkina, G. Z., and Braginskaya, L. L.,** Clinical picture and treatment of chronic poisoning by chlorinated hydrocarbons in workers engaged in the production of trichloroethylene and monochloracetic acid, *Gig. Tr. Prof. Zabol.*, 17, 48, 1973 (in Russian).
397. **Efthymiou, M. L., Vincent, V., and Jouglard, J.,** Les manifestations pulmonaires toxiques enregistrées dans les centres d'informations téléphoniques, *Poumon. Coeur*, 26, 997, 1970.
398. **Milby, T. H.,** Chronic trichloroethylene intoxication, *J. Occup. Med.*, 10, 252, 1968.
399. **Griffiths, W. C., Lipsky, M., Rosner, A., and Martin, H. F.,** Rapid identification of and assessment of damage by inhaled volatile substances in the clinical laboratory, *Clin. Biochem.*, 5, 222, 1972.
400. **Pebay-Peyroula, F.,** Élimination pulmonaire des toxiques-mesure-applications toxicologiques, *J. Eur. Toxicol.*, 3, 300, 1970.
401. **Gobbato, F. and Ceccarelli, S.,** Importanza e limiti dell'esame otofunzionale nella diagnosi precoce di intossicazione professionale, *Folia Med.* (Naples), 54, 93, 1971.
402. **Lacroix, G. and Marozzi, E.,** Sulla diagnosi tossicologica in tema di avvelenamento mortale da sostanze organiche alifatiche clorurate, *Minerva Medicoleg.*, 79, 143, 1959.
403. **Zaffiri, O., Alessio-Verni, A., and Canali, E.,** La rianimazione nelle intossicazioni da prodotti industriali, *Minerva Anestesiol.*, 34, 1117, 1968.
404. **Zaffiri, O., Bisiani, M., Francescato, F., and Del Prete, D.,** Trattamento rianimatorio in un caso di intossicazione acuta da trielina, *Rass. Int. Clin. Ter.*, 46, 1065, 1966.
405. **Nathanzon, G. E., Markin, E. A., and Afans'eva, L. I.,** Clinical picture and therapy of acute oral trichloroethylene poisoning, *Gig. Tr. Prof. Zabol.*, 15, 46, 1971 (in Russian).
406. **Betti, V. and Franchi, L.,** Rianimazione respiratoria d'urgenza nell'intossicazione acuta da tricloroetilene, *Omnia Med. Ther.*, 44, 715, 1966.
407. **Baetjer, A. M.,** Dehydration and susceptibility to toxic chemicals, *Arch. Environ. Health*, 26, 61, 1973.
408. **Roessler, R. and Morawiec-Borowiak, D.,** Zastosowanie hemodializy w ostroym satruciu trójchloroethylenem, *Wiad. Lek.*, 26, 1271, 1973.
409. **Gaultier, M. and Potter, M.,** La réanimation cardiaque au cours des intoxications, *Sem. Ther.*, 43, 190, 1967.
410. **Kubacki, A. and Siekierski, J.,** Vandid w leczeniu zaburzén oddechowych w prezebiegu zatrucia trójchloroetylenem i innymi środkami chemicznymi, *Wiad. Lek.*, 19, 387, 1966.
411. **Bothe, J., Braun, W., and Dönhardt, A.,** Untersuchungen zur antidotwirkung von paraffinöl bei Vergiftungen mit Kohlenwasserstoffen an der Maus, *Arch. Toxikol.*, 30, 243, 1973.
412. **Smyth, H. F.,** Hygienic standards for daily inhalation, *Am. Ind. Hyg. Assoc. J.*, 17, 129, 1956.
413. **Irish, D. D.,** Common chlorinated aliphatic hydrocarbon solvents, *Arch. Environ. Health*, 4, 320, 1962.
414. **Ashe, H. B., Baier, E. J., Coleman, A. L., Elkins, H. B., Grabois, B., Hayes, W. J., Jr., Jacobson, K. H., Reindollar, W. F., Scovill, R. G., Smith, R. G., and Zavon, M. R.,** Threshold limit value for 1963, *J. Occup. Med.*, 5, 491, 1963.
415. American Conference of Governmental Industrial Hygienists, *Documentation of the Threshold Limit Values for Substances in Workroom Air*, 3rd ed., A.C.G.I.H., Cincinnati, 1971, 263.
416. **Utidjian, H. M. D.,** Criteria for a recommended standard. Occupational exposure to trichloroethylene. I. Recommendations for a trichloroethylene standard, *J. Occup. Med.*, 16, 192, 1974.
417. Department of Labor, Occupational Safety and Health Administration, Trichloroethylene: proposed occupational exposure standard (29 CFR Part 1910), *Fed. Regist.*, 40, 49032, October 20, 1975.
418. **Messite, J.,** Trichloroethylene, *J. Occup. Med.*, 16, 194, 1974.
419. **Luxon, S. G.,** Recent developments in the use of solvents, *Ann. Occup. Hyg.*, 9, 231, 1966.
420. **Schaffer, A. W.,** Toxicite comparee des solvants chlores, *Arch. Mal. Prop. Med. Trav. Secur. Soc.*, 31, 150, 1970.
421. **Morgan, D. J.,** Assessment of exposure to trichlorethylene, *Ann. Occup. Hyg.*, 7, 365, 1964.
422. **Ulrich, L. and Rosival, L.,** Kotazke koncepcie najvyssie pripustnych koncentracii chemickych latok v pracovonom prostredi, *Bratisl. Lek. Listy*, 58, 366, 1972.
423. **Weichardt, H., and Lindner, J.,** Messmethoden zur Überprufung des MAK-Wertes an verschiedenen Trichloräthylen-Waschanlagen, *Zentralbl. Arbeitsmed., Arbeitsschutz*, 22, 323, 1972.
424. **Ikeda, M. and Imaura, T.,** Biological half-life of trichloroethylene and tetrachlorethylene in human subjects, *Int. Arch. Arbeitsmed. Arbeitsschutz*, 31, 209, 1973.
425. **Lindner, J.,** Felduntersuchungen in Lösungsmittelbetrieben. 3. Korrelation zwischen Lösungsmittelkonzentrationen (Trichloräthylen) der Arbeitsplatzluft und biologischem Material als Mittel zur Arbeitsplatzuberwachung, *Zentralbl. Arbeitsmed. Arbeitsschutz*, 23, 161, 1973.
426. **Silverman, P.,** Behavioural toxicology, *New Sci.*, 61, 255, 1974.
427. **Schmidt, K. and Zorn, H.,** Diagnostik unklarer Vergiftungsfälle durch Atemluftanalyse, *Dtsch. Med. Wochenschr.*, 97, 1507, 1972.
428. **Williams, J. W.,** The toxicity of trichloroethylene, *J. Occup. Med.*, 1, 549, 1959.

AUTHOR INDEX

D

E

F

G

H

I

J

K

L

M

N

S

T

U

V

W

Y

Z

SUBJECT INDEX

O

P

Q – S

T

U – W